AF553682

CHEMICALS IN THE ENVIRONMENT

Y. MIDO
Dept. of Chemistry
Kobe University
Kobe 657, Japan

M. SATAKE
Faculty of Engineering
Fukui University
Fukui 910, Japan

Consulting Editors

M.S. SETHI
A.R.S.D. College
University of Delhi
New Delhi-INDIA

S.A. IQBAL
Dept. of Chemistry
Saifia P.G. College of
Science and Education
Bhopal, INDIA

Discovery Publishing House
New Delhi (INDIA)

First Published - 1995

Reprinted - 2016

ISBN: 978-81-7141-267-9

Chemicals in the Environment

Published by:

DISCOVERY PUBLISHING HOUSE PVT. LTD.
4383/4B, Ansari Road, Darya Ganj
New Delhi-110 002 (India)
Phone: +91-11-23279245, 43596064-65
Fax: +91-11-23253475
E-mail: discoverypublishinghouse@gmail.com
sales@discoverypublishinggroup.com
web: www.discoverypublishinggroup.com

Printed at:
Infinity Imaging Systems
Delhi

Preface

The teaching of Chemistry at the introductory stage becomes each day a more challenging task as the subject matter becomes more diverse and more complex. These challenges have evoked a series of responses – the present set of introductory chemistry monographs is one such. The teaching of chemistry recognises a number of problems that confront those who select text books. In order to overcome these problems, this volume "**Chemicals in the Environment**" – one of about fifty in the Chemistry Monograph Series – is introduced. Each volume is independent of the others deals with one of Chemistry topics and constitutes a complete entity. Each volume is more comprehensive than can be possible in a single volume text. It is intended to provide a range of topics to cover most undergraduate and chemistry main courses of study. These volumes can be used to enrich the more conventional courses of study.

Suggestions for improvement are welcome and shall be gratefully acknowledged.

Authors

Contents

CHAPTER 1

Introduction

Chemistry and the Biosphere

The natural environment of living things is a complex, though fragile, membrane around the earth. This part of the earth is abundantly rich in the quantity and variety of animals, plants, fungi, bacteria and other living things. All living organisms, together with the physical and chemical component of the total environment, make up the biosphere (often called the ecosphere). It involves all the interaction between the living and non living components that produce, support, and affect life. The biosphere is more than the abode of living things - it is a self-sufficient biological system in which the thread of life is sustained in a dynamic state of equilibrium by a flow of energy from the nearest star, the sun. The energy input from the sun captured, used over and over again by the web of life, and eventually deposited in the earth or emitted back into space.

The biosphere is remarkably constant in physical and chemical characteristics. So special are these conditions that perhaps nowhere are they exactly duplicated on any satellite among all the stars in the Universe. However, there may be many that are similar. The temperatures of bodies in space range over millions of degrees, from intense coldness to the hottest star - the Sun. On the contrary, the biosphere of the planet earth is an island of tranquillity, where the temperature is always within a few degrees of that necessary to maintain water, the giver of life, in its liquid state. The size of the earth and its gravity, rotation, and distance from the sun, as well as the chemical composition of the rocks, the ocean, and the atmosphere, hold within the membrane a delicate balance in which life can flourish. Only slightly beyond these limits, life, as we know it, would perish.

Man's activities in the material sphere have qualitatively different directions. While investigating this phenomenon as a natural event, it is convenient to distinguish two aspects of man's activities; physical (among these, thermodynamic, mechanical, electromagnetic), and chemical and biological.

For the biosphere, different aspects of man's activities are apparently, dangerous at various stages of the development of civilization. Until recent times the *biological* aspect was the most threatening; man annihilated the large animals and herded sheep and goats which literally "consumed" whole regions; he felled forests and thus brought on degradation of the top soil, promoting the advance of the desert. This factor exists even today. However, among man's disruptive activities there arose a new and now dominating aspect, the chemical.

Those of man's activities which directly affect the chemical form of the movement of material in the biosphere may be termed chemical activities. This aspect of man's activities is now the cause of greatest concern. For it is the chemical activity that most clearly shows up in the disruptive effects on the biosphere of 20th century civilization. It is possible that in the future this role will come to belong to the physical factor (primarily, to the thermodynamic, i.e. disruption of the total thermal balance of the planet).

At present man's chemical activities have reached such rates that they surpass nature's capacity to neutralize them with the aid of natural feedback systems (that is, without any intervention by man). Nature can no longer wait for favours from mankind; we are obliged to help Nature.

Increase in Man-made Chemicals

Man's chemical activities which affect nature are. (*a*) The Consumption of chemical substances from nature; (*b*) The mass introduction of substances from the depth of the earth into the environment: (*c*) Contamination of nature by the wastes from man's activities; and

(*d*) The appearance in nature and, in particular, in the biosphere of new, highly active chemical compounds separated out from natural sources or synthesized by man.

More than 6 million chemical compounds obtained from natural sources and via synthesizing procedures, are already known. Two hundred thousands substances - mainly synthetic – are added to the list each year.

Man has created a whole universe of new chemical compounds and is rapidly expanding its boundaries. Each compound has its basic physical, and chemical properties. In the universe of chemical compounds, products which are both beneficial and dangerous to the biosphere, and to man in particular, are concealed.

There are some 70,000 chemicals being traded, of which as many as 35,000 have been classified by the Organization for Economic Co-operation and Development (OECD) and US Environmental Protection Agency (EPA) as definitely or potentially harmful to human health. The chemical industry produces huge profits but also enormous quantities of hazardous wastes. The 'term Toxic' describes a narrow group of substances that cause death or serious injury to human beings or animals, while 'hazardous' is a broader term describing that present immediate or long-term threats to people or the environment.

Most toxic and hazardous chemicals, wastes come from industries that produce plastics, soap, synthetic rubber, fertilizers, synthetic fibres, medicines, detergents, cosmetics, paints, pigments, adhesives, explosives, pesticides, herbicides and chemicals.

Transporting toxic chemical contains an element of risk. In 1985, a PCB leak from a lorry closed down 100 miles of the Trans Canada Highway for a week. IN 1987, after the sinking of the ferry *Herald of Free Enterprises* in Belgium, rescue attempt were endangered by the toxic chemical that were on board—25 drums of toluene di-isocyanate (TDI), which reacts with water to create dangerous fumes; six drums of cyanide; and several other containers of different chemical waste products.

In the UK., the lorries carrying hazardous substances display coded signs so that rescue services can modify their operations accordingly in the event of accident.

From World Health Organization literature, *World Resources 1987* has compiled a list of toxic chemicals—

Arsenic and compounds
Mercury and compounds
Cadmium and compounds
Beryllium and compounds
Thallium and compounds
Chromium (VI) compounds
Lead and compounds
Antimony and compounds
Phenolic compounds
Cyanide compounds
Isocyanates
Organohalogenated compounds, excluding polymeric materials
Chlorinated solvents
Organic solvents
Biocides
Tarry materials from refining
Pharmaceutical compounds
Peroxides chlorates, perchlorates, azides
Ethers
Chemical laboratory materials, not identifiable and/or new, with unknown effects on the environment
Asbestos
Selenium and compounds
Tellurium and compounds
Polycyclic aromatic hydrocarbons
Metal carbonyls

Soluble copper compounds PCBs

Acids and or basic substances used
in the surface treatment and
findishing of metals

The data reveal that all chemical compounds possess one or another type of biological activity - that is, introduced into a living organism, they have some effect on the organism's specific properties and functions. Even seemingly inert substances such as the noble gases (helium, neon, and others) have a strong narcotic effect; having entered the human organism, insoluble, inert particles of asbestos can give rise to a cancerous tumor; fine particles of silicon are capable of effectively routing some insects.

Man has created some 10,000 drugs of synthetic origin, which together with food additives generate their own type of stress, continuously and concentratedly stressing man. As a representative of a distinct biological species, what will man's reaction be to the effects of such substances not previously encountered along the paths of man's biological evolution ? It is not possible to answer this question at the present time. It is obvious that chemistry has amassed a whole arsenal of means of affecting the biosphere without knowledge of their powers and potentialities.

Eugene Odum (1977) has provided data which show the relationship of many respiratory ailments to contamination of the environment. n general, man's activities directed towards changing the chemical potential of the planet have led to the emergence of a whole system of uncontrolled processes, the contents and interconnections of which are not clear in a majority of instances. Having instigated these processes, put our hopes on them, and promoted them, we condemn those consequences which turn out to be hazardous or harmful to us. However, so long as we do not understand the processes we are unable to regulate them but also to forecast their possible consequences.

Cause for Concern

The attainment of chemistry and industry have substantially exceeded the resources of biology and medicine to forecast scientifically the effects of chemical products on the biosphere. Mankind lacks adeauate systems for tracking the chemical composition and state of the biosphere. Incomplete information about chemical changes in the biosphere on the one hand and about the biological activity of chemical compounds on the other leads to a situation in which man is not able to predict distant consequences of changes in the chemical composition of the environment.

It is only in recent times that man has begun to recognize the scale and importance of his industrial (in particular, chemical) activities. Until now man has not been ready, to the required degree, for a conscious, objectively oriented, collective effort toward optimization of this activity.

The problem posed by man's chemical activities and the need for nature preservation is vast and many-sided. Among a great number of aspects and questions are the contamination of rivers by industrial wastes, the extensive and virtually uncontrolled use of chemical fertilizers and herbicides in agriculture. There are scientific problems which are far beyond the present state of man's knowledge such as, the widespread use of antibiotics; there are problems which are obscure insofar as method of implementation e.g., the creation of an adequately diversified and safe complex of contracetive agents.

REFERENCES

Gamow, G. (1970) *A Planet called Earth*. Bantom Books, New York.

Hutchinson, G.E. (1970) The Biosphere. *Scientific American,* 223(3), 45-53.

Jessop, N.M. (1970) *Bisphere. A Study of Life*. Prentrice Hall, Englewood Cliffs. N.J.

Odum, E.P. (1971) *Foundations of Ecology*. W.B. Sounder, Philadelphia.

Southwick, C.H. (1976) *Ecology and the Quality of Our Environment* (2nd Ed). William Grant Press, Bostom.

CHAPTER 2

Assessment of Toxicity

DEFINITIONS

The following terms describe the states of matter in which chemical contaminants may occur in the environment :

Gas : A formless fluid that completely occupies the space of an enclosure at 25°C.

Vapour : The gaseous phase of a material that is liquid or solid at 25°C.

Aerosol : A dispersion of particles of microscopic size in a gaseous medium. these may be solid particles (dust, fume, smoke) or liquid particles (mist, fog).

Dust : Airborne solid particles (an aerosol) that range in size from 0.1 to 50 μ and larger in diameter. A person with normal eyesight can detect dust particles as small as 50 μ i diameter. Smaller airborne particles cannot be detected by unaided eyes unless strong light is reflected from the particles. Dust of respirable size (below 10 μ) cannot be seen without the aid of a microscope.

Fume : An aerosol of solid particles generated by condensation from the gaseous state, generally after volatilization from molten metals. The solid particles that make up a fume are extremely fine, usually less than 1.0 μ in diameter. In most cases, the volatilized solid reacts with oxygen in the air to form an oxide. A common example is cadmium oxide fume.

Smoke : An aerosol of carbon or soot particles less than 0.1 μ in diameter that results from the incomplete combustion of carbonaceous materials such as coal or oil. Smoke generally contains droplets as well as dry particles.

Mist : An aerosol of suspended liquid droplets generated by condensation from the gaseous to the liquid state or by the breaking up of a liquid into a dispersed state, such as by splashing, foaming, or atomizing Mist is formed when a finely divided liquid is suspended in the atmosphere. Examples are the oil mist during cutting and grinding operations, acid mists from electroplating, acid or alkali mists from pickling operations, and paint spray mist from spraying procedures.

Fog : A visible aerosol of a liquid, formed by condensation.

The following terms of measurement are commonly used :

ppm : Parts of vapor or gas per million parts of contaminated air by volume.

mg/m³ : Milligrams of a substance per cubic meter of air.

mppcf: Millions of particles of a particulate per cubic foot of air (mainly historical use).

Toxicology is the study of the effects of poisonous substances on living organisms. Above a certain concentration, toxicants have detrimental effects on some biological function. The concentration at which a significant detrimental effect occurs is determined by the *dose response.* The critical (or threshold) dose at which toxicity occurs differs between species, sexes and individuals within a species due to genetic and other factors such as the composition of the diet and some illnesses.

The **dose**, or degree of expose of an organism to a toxicant, can be expressed as:

(*a*) the amount of toxicant present in the organism (units of mass of toxicant per unit weight of body mass of the organism);

(*b*) the amount of the toxicant entering the organism (in the diet, drinking water, or inhaled air in animals and absorbed through the roots or through the leaf cuticle in the case of plants);

(*c*) the concentration in the environment of the organism (duration of exposure is important).

The effect of the toxicant dose is called the *response*. This can vary from no effect to death. Toxicity is commonly categorized on the basic of the duration of exposure, that is acute, chronic and subchronic.

Acute exposure involves a single dose whereas chronic exposure refers to exposure over a long time period (often almost a lifetime - two years in the case of test rodents). Sub-chronic exposure is dosing over a shorter time period – fraction of a lifetime, such as one eight of an experimental rodent's lifetime.

Acute toxicity is caused by fast poisons which include both synthetic and naturally occurring compounds, such as the botulinum toxins produced by the soil bacterium Clostridium botulinum, the venom of certain snakes (e.g. rattle snake and cobra) or species of spider (e.g. black widow spider) and plant-derived toxins such as strychnine and nicotine and some synthetic chemicals, including organo-phosphorous compounds phosphine(PH_3), phosgene ($COCl_2$) and sodium fluoracetate. These substances are classed as supertoxins because they cause lethal effects in humans at doses of less than 5 mg/kg body weight.

The subchronic effects of chemicals are determined by investigating the biochemical and other changes which take place over a period of months. Investigations on chronic effects will examine effects on the lifespan of the organism, cancer induction, changes in geriatric conditions and effects on the offspring caused by exposure of the parent to toxic chemicals. However, it is important to note that the acute and sub-acute effects of toxins determined in laboratory experiments cannot always be relied on to predict responses to the same chemicals in the environment. This is due to instruction between pollutants (antagonistic and synergistic effects) and reactions of the toxicants with the components of the environment (such as adsorption, photodecomposition, acidification and dissolution). In the case of non-carcingens, it is possible that there may safe or threshold dose levels of toxins (NOAEL = no observed adverse effect level). However, carcinogens are not considered to have safe or threshold concentrations because a single genetic change may lead to an uncontrolled reaction.

Routes of Entry of Chemicals into the Body

Inhalation is the most important route of entry of chemical agents into the body. Next is contact with the skin. In either case, there may be irritation of contaced tissue and/or absorption into the blood with possible systemic intoxication. Although the gastrointestinal tract is a potential site of absorption, the ingestion of significant amounts of chemicals is rare in the industrial situation.

Inhalation : The water solubility of a gas or vapour is an important factor in determining the amount of inhaled material that reaches the lung. Water-soluble gases dissolve readily in the moisture associated with the mucous membrane of the nose and upper respiratory tract. This causes irritation at these sites. At low airborne concentrations, relatively little of these substances will reach the lungs owing to the ''scrubbing'' effect.

At high atmospheric concentrations, however, some of the gas or vapour will not be absorbed at these upper respiratory sites, and amounts sufficient to cause server irritation and pulmonary edema can reach the alveoli.

Comparatively insoluble gases, such as nitrogen dioxide are not removed by the moisture in the nose and upper respiratory tract and can easily reach the terminal recesses of the lung. Substances of intermediate solubility, such as ozone cause irritation of both the upper respiratory tract and the lung.

Gases such as carbon monoxide do not irritate the respiratory tract but are rapidly absorbed into the blood, resulting in systemic intoxication.

The particle size of aerosols determines the extent of their accessibility to small airways. If the diameter of the particles is larger than 10 μ, impaction occurs on the mucous coat of the pharynx or nasal cavity, and the particles do not reach the alveoli. For this reason, particles of 10 μ or less in diameter are termed respirable. Particles between 1 μ and 5 μ often sediment within the bronchioles, whereas particles less than 1 μ can diffuse within the alveoli.

Particulate matter may deposit on the ciliary epithelium, which covers the upper respiratory tract down to the level of he terminal bronchioles. These particles, either in a free state or after phagocytosis, are moved by cilia and the mucous blanket toward the glottis, where they are swallowed or expectorated. Particulate matter that deposits beyond the ciliary epithelium may be absorbed through the alveolar lining into the blood or, as free and phagocytized particles, may enter the lymphatic system. Several industrially important sub-stances, such as crystalline silica, beryllium, and asbestos, resist solubilization by the blood or removal by phagocytes, and remain in the alveoli indefinitely, Irritation, inflammation, fibrosis, allergic sensitization, or malignant change may occur. However, substances such as iron oxide can be present for extended periods of time with no apparent ill effects.

Ingestion. Ingestion occurs as a route of entry through eating or smoking with contaminated hands or in contaminated work areas. Ingestion of inhaled material also occurs. The amounts ingested are usually of little significance. Exceptions are highly toxic substances, such as lead, arsenic and mercury.

Skin Contact. When a substance contacts the skin, four actions are possible

- The skin and its associated film of lipid and sweat may act as an effective barrier that the substance cannot penetrate.
- The substance may react with the skin surface and cause primary irritation.
- The substance may penetrate the skin and cause sensitization.
- The agent may penetrate the skin, enter the blood, and act systemically.

Only a small number of substances are absorbed through the skin in hazardous amounts. In order to pass into the skin, the substance must enter through one or more of the following four routes: the epidermal cells, the sweat glands, the sebaceous glands, or the hair follicles. The pathway through the epidermal cells and the overlaying stratum corneum into the blood is probably the main avenue of penetration, since this tissue constitutes the majority of the surface area of the skin. The stratum corneum plays a critical role in determining cutaneous permeability. Absorption is faster through skin that is abraded or inflamed. For this reason chemicals that are not normally considered hazardous may be dangerous to persons suffering from active inflammatory dermatoses.

The effect of pollutants on animals and plants are listed below.

Effects of Pollutants on Human and Other Mammals

Biochemical effects.

(*a*) impairment of enzyme function by the binding of the toxicant to enzymes, coenzymes, metal activators, or enzyme substrates
(*b*) alternation of cell membrane or carriers in cell membranes.
(*c*) interference with lipid metabolism, resulting in excess lipid accumulation
(*d*) interference with respiration
(*e*) interference with carbohydrate metabolism
(*f*) stopping or interfering with protein biosynthesis through toxic effects on DNA
(*g*) interference with regulatory processes mediated by hormones or enzymes

Clinical response to toxicants :

(*a*) alteralions in the vital signs of temperature, pulse rate, respiratory rate and blood pressure.
(*b*) abnormal skin colour
(*c*) effects on the eye, which include:
   miosis (excessive contraction of pupil)
   mydriasis (excessive pupil dilation)
   conjunctivitis (inflamation of the membrane covering the front of the eyeball)
   nystagmus (involuntary movement of the eyeballs)
(*d*) gastrointestinal effects: pain, vomiting, paralytic ileus (stoppage of normal peristalsis)
(*e*) central nervous system effects: convulsions, paralysis, hallucinations, ataxia and coma

Sub-clinical effects of toxicants

(*a*) damage to the immune system
(*b*) chromosomal abnormalities
(*c*) modification of the functions of lwer enzymes
(*d*) slowing of the conduction of nervous impulses

Teratogenesis, Mutagenesis, Carcinogenesis, and Immune System Defects

Teratogenesis is the creation of birth defect arising from damage to embryonic or foetal cells or from mutations in egg or sperm clls. The biochemical mechanisms of teratogenesis include: enzyme inhibition by xenobiotics, deprivation of essential nutrients and alternation of the placental membrane.

Mutagenesis is the creation of mutations by chemicals or ionizing radiation which bring about alterations to DNA to produce inheritable traits. Although mutations can occur naturally in the absence of xenobiotic* substances, most mutations are harmful. Mechanisms of mutagenicity are similar to those of carcinogenesis and teratogenesis.

Chemical carcinogenesis (or cancer) occurs when xenobiotic substances cause uncontrolled cell replication (i.e. cancer). From the layman's view points, carcino-genesis is the most commonly associated toxie effect of hazardous substances.

Assessment of Short-term Lethality and Acute Toxicity

The most commonly used assessment of toxicity is the measurement of short term lethality. For a given substance, this involves determining the median concentration which is lethal to a fixed proportion, usually 50%, of a test population of organisms after continuous exposure for a fixed time, usually 48 or 96 hrs. This is interpolated from the sigmoidal dose response curve (Fig. 2.1) which is obtained by plotting percentage mortality at each dose against dose.

The lethal concentration (LC) for 50% of a population after continuous exposure for 48h is referred to as the 48h LC_{50} or LD_{50} (LD = lethal dose). This abbreviation may be altered appropriately to correspond to different exposure times and proportions of the population. Usually the concentration referred to is that in aqueous solution but it may also be a concentration in air. If the test compound is insoluble or sparingly soluble in water, it must be uniformly dispersed for a repeatable LC_{50} to be obtained. If an emulsifier or other solubilizing agent is used for this purpose, it should be chosen carefully to ensure a minimal contribution to the toxicity of the system.

Difficulties arise where organisms are exposed to harmful substances in association with particulate matter, or in their prey or other food. What then is the effective concentration to which they are exposed? The nature of the particulate matter will determine whether the substances are

* *Xenobiotic substances are chemical compounds that are foreign to an organism.*

ingested or not. It will also determine the potential for dissociation of toxic substances following ingestion. Normally, only dissociated material can

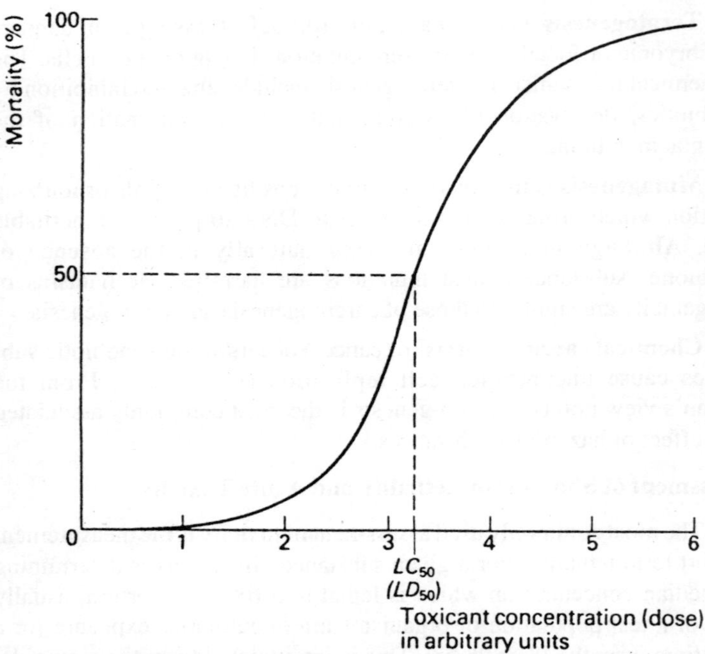

Fig. 2.1 Graph showing typical relationship between mortality and toxicant concentration, or dose, following continuous exposure of organisms for a short fixed time, usually 48 h or 96 h.

exert toxic effects. In a food organism, toxic substances may be converted to derivatives of greater or less toxicity and may be localized in specific parts of the organism, some of which may be selectively eaten or discarded by its predators. Therefore, in these cases it is almost impossible to know the effective concentration to which the predator is exposed. The best one can do is to copy the natural situation as closely as possible in the laboratory and express the LC_{50} as a notional concentration, taken as the total available toxic material divided by the weight of particulate matter or by the weight of the food organism. If one knows that a certain derivative of the toxic substance is much more lethal than any of the others, and suitable methods of analysis are available, the derivative concentration in the food organism should be determined since this will be the most significant factor. In medical usage the LC_{50} frequently refers to an injected concentration, and this must be borne in mind when attempting to extrapolate from such studies.

Lethal concentrations have been expressed in a variety of units, most frequently in milligrams per litre or per kilogram body weight. Where possible, molarity (moles per litre) or molality (moles per kilogram) should be used as this will give a uniform chemical basis for comparing toxicity. The most information the LC_{50} gives is an idea of the order of magnitude of the lethal dose under specified conditions. However, it is relatively quick and cheap to determine and provides a basis, however rough, for initial assessment of the likely hazard from a toxicant and of the effects of various parameters on its toxicity.

Apart from the LC_{50}, a number of other figures may be derived from studies of short term lethality. For example, the minimum lethal dose (*MLD*) is the dose which will kill at least one organism in the test population over the test period. Lucky and Venugopal have proposed that the concept of potential toxicity (p*T*) be introduced on an analogy with pH, i.e. $pT = -\log [T]$ where $[T]$ is the molal concentration of the toxicant and the logarithm is to the base 10. A toxicant with a 24th LD_{50} of 0.001 mol kg^{-1} would have a 24th $[T_{50}]$ of 10^{-3} and a 24h pT_{50} of 3. The calculated p*T* corresponds directly to the effect of toxicant on the test population. A small pT_{50} should indicate a relatively harmless substance while a large pT_{50} will indicate highly toxic substance, at least in the context of lethality in the test system.

Measurement of short term lethality is one aspect of assessment of acute toxicity. Hunter and Smeets, 1977 have defined acute toxicity as 'the total adverse effects produced by a toxicant when administered as a single dose' and this will be the definition applied here. However, alternative definitions are used. e.g. 'the total adverse effects produced by a toxicant when given in a single dose or multiple doses over periods of 24 hours or less'. Therefore, care must be taken to understand the terminology used by a particular author before any appraisal of results can be undertaken. Both definitions quoted above refer to total adverse effects which emphasizes the point that there may be many harmful effects before lethality supervenes, and all effects of the toxicant must be monitored throughout testing. Further, through *post mortem* examinations must be carried out on any organisms that die. Another parameter may be defined. i.e. the EC_{50} or ED_{50}, which is the concentration or dose which produces a specific ill effect in 50% of the organisms tested. The EC_{50} is subject to the same provisos given for the LC_{50}.

An alternative approach to the measurement of acute toxicity is to measure the ET_{50} which is the median exposure time to a given concentration of the toxic substance required to produce 50% mortality in a test population. This is measured range of concentrations and requires continuous monitoring. It therefore involves much more laboratory effort but it can yield more useful information than the 48h LC_{50}. For example, a plot of the In ET_{50} against 1n concentration of toxic substance may indicate a threshold concentration below which the substance is not lethal (Fig. 2.2a). However, it may indicate that the substance is lethal at all concentrations, low concentrations simply requiring long exposures to take effect Fig. 2.2b). Such plot may be strongly suggestive however they cannot be regarded as conclusive evidence. Thus, where there appears to be a threshold in these experiments, (which are of fairly short duration) long exposures may lead to significant mortality due to accumulation of the toxic substance. Similarly, an abrupt threshold may appear in a plot which otherwise displays a continuous trend.

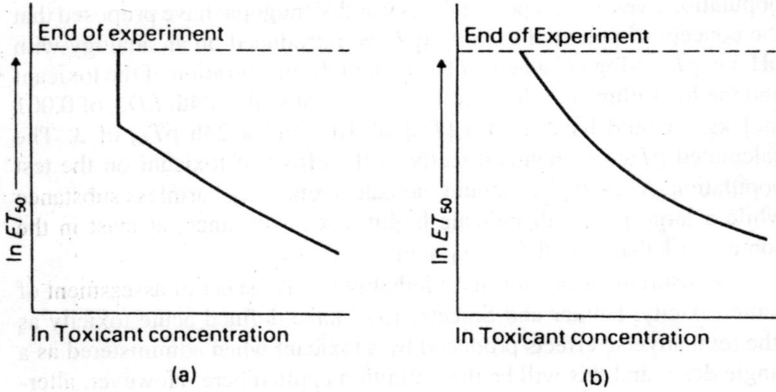

Fig. 2.2 Possible relationship ET_{50} and toxicant concentration: a graph indicating the existence of a safe threshold concentration: b. graph suggesting that the toxicant will be lethal at all concentrations, given sufficient exposure time.

Assessment of Long-term Lethality and Chronic Toxicity

In many environmental situations the observed problems are caused by the response of organisms to concentrations of toxic substances which are harmful only after long periods of continuous exposure. One consequence of such exposure may be death, but the identification of such death as long term lethality from the toxicant is difficult to establish. The death may be due to natural cause, enhanced or accelerated by the weakened state of the test organism but not directly due to the toxicant. Further, the longer the time over which lethality is determined, the greater the variability of the results. All the limitations of the short-term LC_{50} also apply, so long-term lethality is not normally considered a useful parameter for measurement. Chronic toxicity may be defined as 'total adverse effects produced by a toxicant when administered continuously over a long period of time'. Assessing chronic toxicity is not easy, mainly because of the practical difficulties in maintaining organisms under constant conditions in the laboratory for long durations. The fundamental problems are the same as for acute toxicity tests but the time factor greatly exacerbates them. Though almost all laboratory toxicity assessment has been based on continuous exposure of orgainsm to a fixed concentration of toxicant, it is highly unlikely that this situation ever occurs in nature. Therefore, there is desirable set up experimental systems where toxicant concentration fluctuates in a controlled manner, particularly where chronic and sublethal effects are being assessed. However, the difficulties in setting up such systems will be considerable.

Practical Problems in Assessing Toxicity

General considerations

The first practical point to be considered in carrying out toxicity tests is the nature of the test population.

Ideally, the organisms used should be genetically identical and pathogen free. They should be kept under sterile conditions in a constant environment and lit with artificial light, as nearly as possible equivalent to sunlight, or a twelve hour day. These are the requirements of the toxicologist whose main concern is a thoroughly repeatable assay of toxicity. Whether these should be prime requirements of an environmental toxicologist is open to question. By definition, the environmental toxicologist is concerned with the risk to natural populations which are genetically heterogeneous, subject to the effects of pathogens and living in a variable environment. It may be that one important effect of a toxic substance will be to climate selectively sensitive individuals from a population. Another effect may be too eliminate individual affected by a specific pathogen. Other effects may only become apparent under extreme environmental conditions. There are many possibilities, none of which are covered by the analytical toxicologist's ideal population. This is not to say that the carefully selected and maintained populations of organisms, such as inbred rate and mice, which have been developed by medially-orientated analytical toxicologists over the years, do not have their place in environmental toxicology. They are useful for the bio-assay of minute amounts of toxicants, especially where the chemistry of these substances is unknown or does not permit sufficiently sensitive chemical analysis. They may also give pointers to effects on other organisms. Direct extrapolation cannot be justified because of the known magnitude of species and generic differences.

Selection of organism and their maintenance

In selecting organisms for toxicity assessment, the environmental toxicologist must bear in mind the relevance of the organisms to the environment of interest. The effects of air pollution on plants may be monitored using lichens, which show high sensitivity to many toxicants. Other plants may be used where appropriate, a limiting factor frequently being case of maintenance in the laboratory. This limiting factor is particularly important in selecting animals for testing. Another important limiting factor is ignorance of the normal characteristics and degree of variability of many organisms. There are very few organisms for which there is sufficient fundamental knowledge relating to their life under controlled laboratory conditions, far less under the conditions of their normal habitat.

Assuming that a suitable organism can be found, i.e. of demonstrable importance in the environment of interest, with a good background of previous study and easily maintained in the laboratory, a breeding population under suitable defined conditions can be established. A breeding population is essential because susceptibility to toxic substances alters with the developmental stage, typically being greatest in embryonic and larval stages, and in senescence. Suitable defined conditions mean that, until the direct toxicity on the organisms has been assessed, the population should be kept pathogen free as far as possible. Adequate nutrition of known composition must be supplied. In most cases this will be maximal nutrition, with food intake being controlled only by the organism. Anything less will lead to competition between individuals for the available food resulting in a high degree of variation in food intake throughout the population. Maximal food intake is not the same as optimal food intake and it is certainly not the most usual occurrence in any natural environment. However, it does provide a defined base from which to work Other environmental factors to be considered are light, temperature, humidity, daylength, nature of living space (which may have behavioural effects), subjection to handling, noise or other disturbances, and, for aquatic organisms, pH, salinity, degree of oxygenation and other parameters of water quality.

Initially, these factors will be established primarily to suit the experimenter's convenience, but the long-term aim must always be to approximate as closely as possible to the particular environment under consideration. This ultimately requires complex, and expensive, control systems and, possibly, an even more complex computer program to analyse the results obtained. At this point, the laboratory experimenter would be wise to consider the possibilities of a properly planned field study. If a suitable area has already been contaminated with the toxic substance of interest, a study in depth should be carried out, monitoring all possible parameters, and the results should be correlated with those from the laboratory.

CHAPTER 3

Atmospheric Toxicants

Many substances in the atmosphere may be toxic. In this chapter the discussion is confined almost entirely to those that are commonly regarded as air pollutants, i.e. carbon monoxide, hydrocarbons, sulphur oxides, particulates and nitrogen oxides. All these toxicants are not equally harmful and, in assessing the risks they pose, allowance must be made for their varying degrees of toxicity. An attempt has been made to do this by assigning effect factors to the toxicants mentioned. The effect factor is defined as the ratio of toxicant LC_{50} to carbon monoxide LC_{50}. Total toxicant emitted to the atmosphere is multiplied by its effect factor to give a new datum, the emission effect. In terms of emission effects, the hydrocarbons are the most serious atmospheric toxicants. The atmospheric toxicants considered in this chapter are largely products of the internal combustion engine.

Attempts to regulate the emission of atmospheric toxicants have necessitated the determination of safe atmospheric concentrations. In relation to the industrial environment, these may be established by reference to threshold limiting values (TLVs), formerly known as maximum allowable concentrations, (MACs). The threshold limiting value of a toxicant, or mixture of toxicants, is defined as the maximum concentration to which it is believed healthy workers may be repeatedly exposed without ill effect, on the basis of an eight hour working day. In relation to the atmosphere, the threshold limiting value is replaced as a reference point by the 'three minute mean concentration'. This is defined as the maximum permissible concentration of a toxicant, or mixture of toxicants, averaged over three minutes, under the conditions most likely to favour high concentrations at ground level beside a stationary source. The three minute mean concentration is often calculated as a fraction of the threshold limiting value, usually one-thirtieth or one-fortieth. Where there is a risk of accumulation of a toxicant, e.g. a heavy metal an absolute mass emission limit may be established. The absolute mass emission limit is the total amount to toxicant that may be emitted from a stationary source per hour, and which may not legally be exceeded.

In this chapter, atmospheric toxicants will be considered individually under separate headings. Their interactions are also considered since these interactions are often the direct cause of harmful effects.

Carbon Monoxide

Sources and Sinks

Carbon monoxide is the most abundant and widely distributed of the toxicants described in this chapter. It is a colourless, odourless and tasteless gas, slightly lighter than air and only very slightly soluble in water. It can burn but does not itself support combustion.

On an annual basis, about ten times as much carbon monoxide enters the atmosphere from natural sources as from transportation and other human activities. Most of this (77.6%) comes from atmospheric oxidation of methane arising from the breakdown of organic residues, especially in tropical regions. Smaller contributions come from the oceans (3.9%) and from the growth and breakdown of chlorophyll in plants (2.6%). This naturally generated carbon monoxide does not pose any serious problems because the rate of emission at any given place is relatively low. However, man-made carbon monoxide tends to be released rapidly in urban areas. This results in localized concentrations 50 to 100 times higher than the global average.

The concentration of carbon monoxide in the atmosphere at any time depends not only on the rate of production but also on the rate of removal. This occurs mainly in the soil, though a small amount is oxidized to carbon dioxide in the lower atmosphere. in the soil, fourteen species of fungi have been identified as the active agents oxidizing carbon monoxide to carbon dioxide. The activity of these fungi varies with the type of soil, being least active in desert soils and most active in the tropics. Cultivated soils are less satisfactory to these fungi than are those covered with natural vegetation. The total capacity of soil fungi for carbon monoxide oxidation seems to be more than adequate to deal with the carbon monoxide produced on a global basis. Unfortunatley, the urban areas, where most anthropogenic carbon monoxide is released, have among the poorest soil reservoirs of these fungi which, therefore, cannot make an effective contribution to lowering the localized high concentrations in these areas.

Toxic effects

Carbon monoxide has little effect on plants and micro-organisms at the maximum concentrations (about 15 ppm) which normally occur, even in urban areas. However, high concentrations (100 ppm and above) are lethal to many animals. The carbon monoxide combines with haemoglobin forming carboxyhaemoglobin, thus reducing the ability of the blood to carry oxygen. There is no doubt that levels of carboxyhaemoglobin above 5% of the total haemoglobin have harmful effects, ranging from headache, fatigue and drowsiness, to coma, respiratory failure and death. It is probably that levels as low as 1% can adversely affect physiological function. For human beings, the equilibrium percentage of carboxyhaemoglobin in the blood during continuous exposure to an ambient-air carbon monoxide concentration of less than 100 ppm can be estimated from the following equation :

$$\% \text{ COHb in blood} = 0.16 \times \text{concentration of CO in air (ppm)} + 0.5$$

where COHb = carboxyhaemoglobin

0.5 = normal background percentage COHb in blood.

Thus, the blood concentration is directly related to the ambient carbon monoxide concentration. The rate of equilibration varies with the physical activity of the exposed person, increasing as this becomes more strenuous. As far as human beings are concerned. allowance must also be made for voluntary intake of carbon monoxide from smoking. This increases the background blood content of carboxyhaemoglobin 2 to 4 times above the level in non-smokers.

Control

The main source of anthropogenic carbon monoxide is the internal combustion engine and so it has become the prime target for control procedures. Since other toxicants are simultaneously released, there is always the possibility that a method of lowering the concentration of carbon monoxide may increase the concentration of these. For example, supplying a stoichiometric air/fuel mixture to the internal combustion engine will give low carbon monoxide and hydrocarbon emissions, but high emissions of nitrogen oxides. The principal technique evolved to reduce carbon monoxide emission has been the development of exhaust system reactors, which convert carbon monoxide to carbon dioxide, and hydrocarbons to carbon dioxide and water. There are two types of reactor : those that are catalytic and depend on platinum or palladium, or a mixture of both, and those that use a high temperature chamber. None of the catalysts available in entirely satisfactory. They are all poisoned by lead and sulphur compounds which may be present in gasoline, and the ethylene dibromide which is added to leaded gasolines to prevent build-up of lead deposits in engines. They are also poisoned by organothiophosphate compounds released from lubrication oil. It is even possible that platinum and palladium released from catalytic reactors into the environment may prove to be toxic. Certainly, water soluble platinum slats have harmful effects. High temperature chambers do not have any of these shortcomings but they are more expensive to operate because they increase fuel consumption. Other approaches to reducing carbon monoxide emission include improvements in engine design, development of substitute fuels and development of new power sources. None of these has so far proved entirely satisfactory.

Nitrogen Oxides

Sources and Sinks

The nitrogen oxides found in the atmosphere are nitrous oxide (N_2O) nitric oxide (NO) and nitrogen dioxide (NO_2).

Nitrous oxide is colourless, with a slightly sweet taste and smell. Its toxic effects are minimal. Nitrous oxide (N_2O) is a stable long-lived gas formed naturally by blue-green algae and by *Rhizobium* bacteria active in the nodules of peas, beans and other legumes. The internal combustion engine, power stations and nitrate fertilizers now produced almost half the amounts arising for the natural sources.

Nitric oxide, NO, is colourless and odorless, and has appreciable toxicity. Nitrogen dioxide, NO_2, is a reddish-brown gas with a choking smell. Like nitric oxide, it has considerable toxicity.

In general, more nitrogen oxides come from natural sources than form human sources. However, nearly all the nitrogen dioxide is anthropogenic.

Microbial decomposition of organic compounds in the soil is the main source of both nitrous oxide and nitric oxide. Burning of fossil fuels is the principal source of anthropogenic nitrogen oxides. The ratio of nitric oxide to nitrogen dioxide emitted varies with the nature of the combustion process, but the total amount of nitric oxide is always greater than that of the dioxide.

In addition to the greenhouse effect, NO_x are also responsible for depletion of the ozone layer ; and production of acid rain.

In the atmosphere, nitric oxide and nitrogen dioxide become involved in the nitrogen dioxide photolytic cycle (Fig. 3.1.). Firstly, nitrogen dioxide molecules split under the influence of ultraviolet light to nitric oxide and atomic oxygen. The atomic oxygen reacts with molecular oxygen to form ozone. The ozone reacts with nitric oxide to give nitrogen dioxide and molecular oxygen. Thus the cycle is complete. As outlined this cycle produces no net change but if hydrocarbons are present, as is usually the case, they interact with nitrogen oxides and ozone and unbalance the cycle. This leads to accumulation of various toxicants, including the peroxyacyl nitrates.

Whether or not nitrogen oxides are removed from the atmosphere by so:l micro-organisms is still unclear. However, there is no doubt that their removal from the atmosphere involves conversion to nitric acid, possible by a reaction with ozone. . This nitric acid is carried to the earth's surface as nitrate salts in rainfall or dust.

Toxic effects

Damage to plans from high levels of nitrogen dioxide in the air has been observed near factories producing nitric acid. In laboratory experiments, nitric oxide has been shown to interfere with photosynthesis in beans and tomatoes, while nitrogen dioxide has been shown to cause necrosis in leaves of cotton, pinto been and endive plants. These effects were observed at atmospheric concentrations of between 1 and 10 ppm, i.e. higher than the normal concentration which rarely exceeds 1 ppm for nitric oxide or 0.25 ppm for nitrogen dioxide even in the worst affected areas.

Studies of acute lethal toxicity in animals show that nitrogen dioxide is about four time more toxic than nitric oxide. Further, there is little evidence that the concentrations of nitric oxide found in the atmosphere constitute a health hazard. Even proven effects of nitrogen dioxide are associated with concentrations higher than those so far detected in the atmosphere. Such effects are, as is frequently the case with atmospheric toxicants, largely localized in the respiratory tract. Nasal and eye irritation is followed by increasing difficulty in breathing, pulmonary oedema and death. Although nitrogen dioxide may never reach a high enough atmospheric concentration for such effects to be detectable, lower concentrations may be sufficient to sensitize organisms to other noxious compounds or to affect certain sensitive individuals.

Control

Most of the methods employed to reduce anthropogenic emission of nitrogen oxides involve modification of the combustion processes involved.

Emissions from vehicles. Much can be done through economies of consumption of motor fuel and by suitable modifying existing engines. The possible measures include :

1. Keeping the compression ratio below the usual 10 : 1. This has the effect of reducing the combustion temperature which is unfavourable to the initial fixation reaction $N_2 + O_2 \rightarrow 2NO$

And hence reduces the yield of NO.

2. Enriching the mixture of fuel to bring the post-combustion level as low as possible. However, this is energy inefficient and leads to an increase in the release of unburnt fuel.
3. No and unburnt fuel can be removed from the exhaust using a catalytic converter—the option adopted in North America. The action takes place in two stages. In the first NO_x is reduced to N_2 using unburnt fuel and CO as reducing agents and in the second stage the remaining CO and fuel are oxidized by injection of air at 400° C. The catalysts are platinum or alloys of platinum/rhodium which are poisoned by leaded fuels. SO_2 in the emissions is converted to SO_3 and thence to sulphuric acid.

Emissions from power stations. Nitrogen compounds are present at low concentrations in the fuels and give rise to some NO_x on combustion, but the principal source is fixation of nitrogen by reaction of air in the furnace.

Sulphur Oxides

Source and Sinks

The sulphur oxides are sulphur dioxide (SO_2) and sulphur trioxide (SO_3), usually emitted in a mass ratio of about 100 to 1. Sulphur dioxide is a colourless, nonflammable gas under normal environmental conditions, and has an irritating smell at concentrations about 3 ppm. Sulphur trioxide is not normally found in the atmosphere because it reacts rapidly with water to form sulphuric acid. The sulphur oxides released to the atmosphere come almost entirely from human activities, especially coal burning, though small amounts come from burning of fuel oil and smelting of sulphide ores. More are produced in the atmosphere by the oxidation of hydrogen sulphide. This contributes up to 57% of the total production of all sulphur oxides. The prime source of atmospheric hydrogen sulphide is the decay of organic matter. Industrial activities and volcanic activity make small contributions. The main oxidant involved may be ozone.

The content of sulphur in coal supplies normally varies between 0.5—4.0% with an average amount of about 1.3% ; of this about 40% occurs as iron pyrites and the rest is organically combined. There may be significant minor inclusions of chlorine in the range of 0.1-0.7% and trances of fluorine, phosphorus, lead and arsenic.

The biogenetic degradation of natural peptides during the formation of petroleum leads to the inclusion of nitrogen and sulphur compounds. It is also possible that some petroleums are of abiogenic origin and originate in the polymerization of methane which is commonly trapped in sediments and elsewhere in the earth's crust.

In contrast to CO_2, SO_2 is highly water soluble (23 g/100 g at 0°C ; 11 g/100 g at 20° C and NTP) and has a lifetime of only a few hours in the atmosphere before it dissolves in surface water. Its release from vehicles is small compared to that of NO_x but power stations are again implicated being responsible for 60% of the SO_2, twice the quantity produced by industry including the refineries.

Unlike nitrogen, sulphur reacts readily with oxygen during combustion :

$$S + O_2 \rightarrow SO_2 \qquad (1)$$

$$SO_2 + \frac{1}{2}O_2 \rightarrow SO_3 \qquad (2)$$

The oxidation to sulphur trioxide (equation 2) does not occur readily in dry air but proceeds when catalysed by platinum, vanadium or sunlight ; ozone is also an effective oxidant (Wayne, 1991). Sulphur dioxide forms sulphuric acid in its reaction with nitric oxide :

$$H_2O + NO + SO_2 \rightarrow H_2SO_4 + N_2O \qquad (3)$$

but in polluted air the principal route involves the OH radical. In the presence of water and oxygen this can be represented as :

$$OH + SO_2 (+ O_2, H_2O) \rightarrow H_2SO_4 + OOH \qquad (4)$$

or more particularly as the initial formation of the bisulphite radical :

$$^{\cdot}OH + SO_2 + M \rightarrow HO\,.\,SO_2 + M \qquad (5)$$

followed by its reaction with water and oxygen.

Much of the atmospheric sulphur dioxide is ultimately oxidized to sulphur trioxide, which is converted to sulphuric acid as a described above. The sulphuric acid then forms sulphates which settle out of the atmosphere or are washed out by rain. The oxidation of sulphur dioxide in the atmosphere is rapid. Catalytic oxidation may occur in water droplets or on the surface of solid particles. In water droplets, the oxidation is thought to involve molecular oxygen, with iron and manganese salts as catalysts. The iron and manganese salts are largely derived from the fly ash of burning coal. The ash particles may also act as nucleation sites for water droplet formation. In addition, photochemical oxidation may occur with ozone or peroxyacyl nitrates in photochemical smog. The result of either of these processes is a mist of sulphuric acid droplets. Catalytic oxidation stops when the sulphuric acid concentration in the droplets reaches one molar. Possible because of the resultant reduction in sulphur dioxide solubility. Neutralization of the acid by metal oxides or ammonia increases the solubility of sulphur dioxide and permits catalytic oxidation to resume. Ammonia is probably particularly important in this process because it occurs naturally in the atmosphere. The solubility of ammonia increases with decreasing temperature and, therefore, its effect may be especially marked at high altitudes where temperatures are low.

Sulphur dioxide and the sulphuric acid formed from it have four adverse environmental effects :

1. toxicity to humans ;
2. acidification of lakes and surface waters ;
3. damage to trees and crops ;
4. damage to buildings.

Toxic Effects

Chronic exposure of plants to low concentrations (about 0.01 ppm) of sulphur dioxide causes leaf yellowing as a result of inhibition of chlorophyll synthesis. This is associated with accumulation of sulphate in the leaves. High concentrations of sulphate (about 0.5 ppm and above) cause rapid leaf destruction. As always, response varies from species to species and with environmental conditions, most notably the presence of other pollutants. For example, ozone and nitrogen dioxide have been shown to exert synergistic effects with sulphur dioxide. The sulphuric acid mists associated with high sulphur dioxide concentrations in the atmosphere are another major source of damage, again primarily through effects on leaves.

Toxicity to humans. The detection limit of SO_2 by humans is about 0.5 ppm and at leaves below 1 ppm it has no obvious effect but breathing difficulties are experienced above 1.5 ppm. Exposure at 200 ppm for 1 minute causes great discomfort, while the limit for prolonged exposure is 5 ppm. The gas exerts its worst effects on asthmatics and bronchitics who are at risk if exposed for 1 day at the detection limit.

Inhalation of SO_2 was established as the cause of death in a number of incidents :

1930 In the Meuse valley levels over 10 ppm built up during a thermal inversion with input from power stations, iron and steel works and other industries. Sixty people died on two December days from heat failure linked to respiratory disorder.

1948 Under similar conditions at Donora near Pittsburg over 10 000 people were affected during 5 days in October and 20 deaths were attributed to stress from SO_2.

1952 The December 5-9, 1952, London fog is the best known air pollution episode in England. It began on Thursday, December 4, as a high-pressure air mass created a temperature inversion over southern England. A white fog formed in the London area. As particulate and sulphur dioxide levels built up, because of

> extensive use of coal as fuel for space heating and electric production, the fog became a black fog. At the same time the high-pressure area stalled and became stationary. The buildup of pollutants combined with the fog resulted in essentially zero visibility. By Saturday pollutants were sufficiently concentrated to cause deaths, which were estimated to be about 4000.

Acute sulphur dioxide effects on animals are mainly related to the respiratory system. A concentration of 1.6 ppm will cause reversible bronchiolar constriction, and detectable respiratory effects increase above this level. Below 25 ppm, irritant effect occur mainly in the upper respiratory tract and the eyes, because in the first tissue water with which it comes in contact. Studies of the effects of chronic exposure to low levels of sulphur dioxide show an increased incidence of respiratory infection. This must be a particular hazard for those already suffering from respiratory deficiencies. Sulphur dioxide dissolved in tissue water is converted oxidatively to sulphuric acid and sulphates as previously described. Sulphates are much more powerful irritants than sulphur dioxide itself and the problem is enhanced by the presence of sulphate aerosols in the atmosphere wherever sulphur dioxide is produced. Concentrations of sulphate as low as 0.002 ppm have been shown to have harmful effects on susceptible people, e.g. asthmatics, and it is not uncommon for concentrations in the atmosphere of this order to be reached or exceeded. An unfortunate consequence of the introduction of catalytic units in vehicle exhaust systems, to remove carbon monoxide and hydrocarbons, has been an increased emission of sulphate, since the catalysts convert the trace amounts of sulphur oxides present into sulphuric acid. Though the total produced is not great, it may well be that the hazard from increased sulphate emissions exceeds the benefits from reduced carbon monoxide and hydrocarbon levels.

Control

Modifying the catalytic units in car exhaust systems to reduce sulphuric acid production may involve changing the method of catalyst use, adding a sulphate trap, or completely removing sulphur from gasoline prior to its use. As far as coal burning, the main source of anthropogenic sulphur oxides, is concerned, there are three possible ways of reducing emission. Firstly, one can use coal with a very low sulphur content. Secondly, sulphur can be removed from coal by a water washing process. However, this is expensive and removes only pyritic sylphur. Thirdly, coal can be gasified and the sulphur oxides produced trapped by a suitable chemical reaction.

Precombustion removal of sulphur is chemically feasible for oil through the pressurized hydrogenation of the fuel and alkaline trapping of liberated hydrogen sulphide :

$$R-S-R+2H_2 \rightarrow 2R-H+H_2S \quad (6)$$

However, the costs are high and are estimated to add 20% to the price of electricity.

There has been considerable research into the cleaning of coal and water washing dissolves part of the iron pyrites and can reduce SO_2 emissions by 10%. The chemical methods largely depend on oxidation and subsequent neutralization of sulphuric acid with lime :

$$4FeS_2+15O_2+8H_2O \rightarrow 2Fe_2O_3+8H_2SO_4 \quad (7)$$

The aqueous slurry is pressurized with oxygen in an autoclave and heated to 150-200° C when most of the inorganic sulphur is removed (equation 7).

Organic sulphur compounds include thiols (RSH) which react completely under these conditions and some alkyl thioethers (RSR) which react in part, but of the remaining known compounds, aryl ethers (PhSPh) and sulphur heterocycles are unaffected. An added disadvantage is the loss of heating value of the order of 10-20% for only 10-20% sulphur removal.

Flue gas desulphurization (FGD). The scrubbing plant injects limestone slurry which converts SO_2 into calcium sulphite :

$$CaCO_3+SO_2 \rightarrow CaSO_3+CO_2 \quad (8)$$

and in the same process air is also injected to produce the more tractable calcium sulphate (gypsum) :

$$CaSO_3+\frac{1}{2}O_2 \rightarrow CaSO_4 \quad (9)$$

Major new environmental problems may arise due to extraction of lime stone.

Fluidized bed combustion (FBC). This technique depends upon the capture of SO_2 by lime or limestone to form calcium sulphate. Coal in the furnace bed in mixed with limestone to form calcium sulphate. Coal in the furnace bed in mixed with limestone and fluidized by passing of air (cf. equation 5.30) at 800-1000 C. A higher temperature must be avoided for two reasons :

1. Calcium sulphate begins to decompose and release SO_2 :

$$CaSO_4 \xrightarrow{\text{c. } 1200^\circ \text{ C}} CaO+SO_2+\frac{1}{2}O_2 \quad (10)$$

2. In the region of 1400° C thermal fixation of nitrogen occurs to produce NO_x ; this may be 40% of the total.

Ozone

Ozone, O_3, is a gas and is a secondary pollutant. It is present in significant amounts in both the troposphere and the stratosphere, the combined amount being known as total column ozone.

Stratospheric ozone is generated by the actor of ultraviolet light on oxygen molecules :

Absorption of higher energy radiation at shorter wavelength, typically at 200 nm, leads to cleavage of the oxygen molecule and the formation of very reactive atomic oxygen (equation 11). When atomic and molecular oxygen collide ozone is formed and survives the collision provided that a third inert entity (M), normally nitrogen present in excess, is available to take up excess energy.

$$O_2 \xrightarrow{hv} [O] + [O] \qquad (11)$$

$$[O] + O_2 + M \rightarrow O_3 + M \qquad (12)$$

$$O_3 \xrightarrow{hv} O_2 + O \qquad (13)$$

Ozone in the Troposphere

The filtering action of the ozone layer removes the shorter wavelength radiation with sufficient energy to produce odd oxygen. Under natural conditions small amounts of ozone enter by transfer from the stratosphere. However, tests in pollution zones reveal substantial levels which in sunlit urban areas can rise to 1 ppm a typical concentration is about 0.3 ppm. In these areas nitrogen is fixed as nitric oxide in the internal combustion engine where oxygen is present in limited excess ($N_2 + O_2 \rightarrow 2NO$. In daylight this initial product reacts with atomic oxygen, or other oxidizing agent present in low concentration in the natural environment, to give the dioxide :

$$NO + O \rightarrow NO_2 \qquad (14)$$

This is the primary source of ozone because it can be photolysed by available blue light of wavelength near 400 nm. So through the morning, sunlight initiates decomposition of the peroxide to give the lower oxide and release atomic oxygen :

$$NO_2 \rightarrow NO + [O] \qquad (15)$$

This then combines with molecular oxygen to produce ozone.

In the absence of other agents, oxygen and $(NO)_x$ would achieve a steady state including a low level of ozone, but equation 15 is driven to the right by the removal of NO through its reaction with other pollutants, notably carbon monoxide and unburnt hydrocarbons. As a result the concentration of atomic oxygen and hence of ozone is further increased.

Diurnal variations of ozone levels

In the stratosphere little change occurs during the dark hours since, although production ceases, the concentration of atomic oxygen falls away and ozone is no longer lost by the reaction :

$$O + O_3 \rightarrow 2\,O_2 \qquad (16)$$

In contrast, levels in the troposphere decline after dark because ozone continues to react with other residual pollutants such as hydrocarbons.

Ozone differs from molecular oxygen in that it is a polarized molecule and hence is twelve times as soluble in water as oxygen under comparable conditions of temperature and pressure : ozone is therefore scrubbed from polluted air by rainfall.

Toxic Effects

The primary site of injury is the lung which may become congested with swelling of the tissue (oedema) and possible haemorrhage. In man exposure for 2 hours to concentrations of 1.5 ppm is regarded as dangerous. Workers particularly at risk are welders using the shielded arc process.

Hydrocarbons and Photochemical Derivatives

Sources and Sinks

Hydrocarbons can exist as gases, liquids or solids under normal environmental conditions. Under such conditions, molecules containing four carbon atoms or less are gases, while those with five or more are liquids or solids. Most of the hydrocarbons contributing to air pollution have twelve or less carbon atoms per molecule and, therefore, are either gases or volatile liquids. These hydrocarbons may be either aliphatic or aromatic. As with carbon monoxide, most hydrocarbons enter the atmosphere from natural sources.

Methane, the simplest hydrocarbon, make the largest contribution on a global basis. Plants emit the more complex hydrocarbons, terpenes and hemiterpenes, the total production of which is about half that of methane. Human activities contribute only about 15% of the total annual release of hydrocarbons to the atmosphere but, as with carbon monoxide, this release, mostly from transportation, is highly localized in urban areas.

The toxic effects of hydrocarbons in the atmosphere are mainly due to derivatives produced by photochemical oxidation (Fig. 3.1).

Methane is an important greenhouse gas. It occurs at 1.7 ppm in air and in addition to losses from processing natural gas it arises from the following biological sources :

(*a*) cultivation of rice ;

(*b*) anaerobic fermentation if swamps and rain forests ;

(*c*) in the intestines of animals, chiefly cattle.

Its effect on surface warming is enhanced by a factor of seven compared to carbon dioxide because it absorbs infrared radiation near 3000 cm^{-1}, which coincides with a window in the spectrum of carbon dioxide.

Reactions in the atmosphere : The hydroxyl radical features critically in the chemistry of air pollution ; it is formed when ozone is photolysed afford excited oxygen which combines with water vapour

$$O^{\cdot} + H_2 O \rightarrow HO^{\cdot} + .OH \tag{17}$$

The $^{\cdot}OH$ radical reacts with CO in clean air to form CO_2 (equation 18) and also abstracts $H^{\cdot}$ atom from methane in polluted air (equation 19) so converting it into the reactive methyl radical :

$$HO^{\cdot} + CO \rightarrow H^{\cdot} + CO_2 \tag{18}$$

$$HO^{\cdot} + CH_4 \rightarrow {}^{\cdot}CH_3 + H_2 O \tag{19}$$

This in turn captures oxygen to give the methylperoxy radical. The natural outcome is the interaction between peroxy radicals to regenerate oxygen, but NO in polluted air diverts the system according to equation 20 :

$$CH_3 O - O^{\cdot} + NO \rightarrow CH_3 O^{\cdot} + NO_2 \cdot \tag{20}$$

Higher alkanes : These reactions are of the same type, for example ethane is converted into acetaldehyde which like formaldehyde, is a volatile sub-stance which also contributes to smog. The further reactions of acetaldehyde are particularly significant since they lead to the formation of peroxyacetylnitrate (PAN), one of the principal irritants in photochemical smog. In the first step (equation 6.6) a relatively stable acetyl radical is formed from acetaldehyde by abstraction of an H atom :

$$\begin{array}{l} \quad\;\; \curvearrowleft {}^{\cdot}OH \\ H \\ \quad \backslash \\ \quad\; C = O \longrightarrow CH_3 - C = O + H_2 O \\ \quad / \\ CH_3 \end{array} \tag{21}$$

After capture of oxygen, which is available in high concentration, the acetylperoxy radical is obtained :

$$CH_3 - \dot{C} = O + O_2 \longrightarrow CH_3 - \underset{}{C}(=O) - O - \dot{O} \quad (22)$$

Notice that NO_2 is itself an odd electron species and pairs to give PAN :

$$CH_3 - C(=O) - O - \dot{O} + NO_2 \longrightarrow O_2N - O - O - C(=O) - CH_3 \ \text{(PAN)} \quad (23)$$

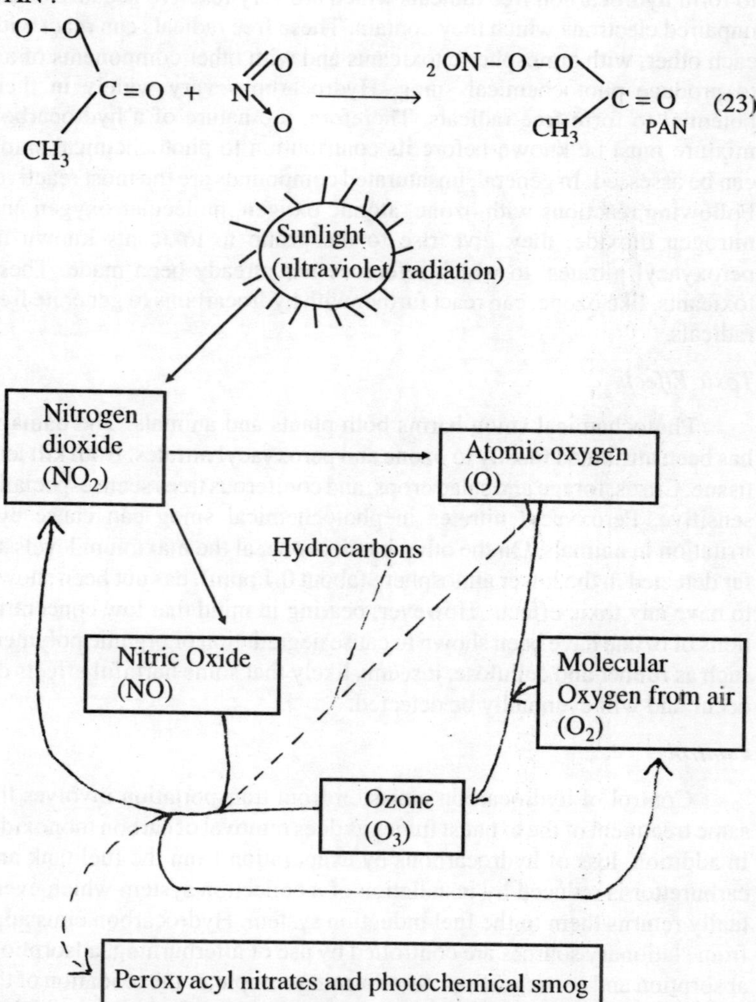

Fig. 3.1. The nitrogen dioxide photolytic cycle, showing the interaction of hydrocarbons to produce photolytic smog.

Such toxic derivatives are sometimes referred to as secondary pollutants. Photochemical oxidation frequently involves ozone and peroxyacyl nitrates, which have the general formula $RCO_3 NO_2$ where R is a hydrocarbon group. Ozone is produced in the atmosphere by the action of sunlight on nitrogen dioxide, which splits it into nitric oxide and atomic oxygen. Each atom of oxygen can then react with a molecule of oxygen to give ozone. If hydrocarbons are present, they can react with the ozone to form hydrocarbon free radicals which are very reactive because of the unpaired electrons which they contain. These free radicals can react with each other, with atmospheric toxicants and with other components of air to produce photochemical smog. Hydrocarbons vary widely in their potential to form free radicals. Therefore, the nature of a hydrocarbon mixture must be known before its contribution to photochemical smog can be assessed. In general, unsaturated compounds are the most reactive. Following reactions with ozone, atomic oxygen, molecular oxygen and nitrogen dioxide, they give rise to the group to toxicants known as peroxyacyl nitrates, to which reference has already been made. These toxicants, like ozone, can react further with hydrocarbons to generate free radicals.

Toxic Effects

Photochemical smog harms both plants and animals. The damage has been attributed mostly to ozone and peroxyacyl nitrates. Both kill leaf tissue. Citrus, forage and salad crops, and coniferous trees seem especially sensitive. Peroxyacyl nitrates in photochemical smog can cause eye irritation in animals. On the other hand, ozone, at the maximum levels so far detected in the lower atmosphere (about 0.1 ppm), has not been shown to have any toxic effects. However, bearing in mind that low concentrations of ozone have been shown to cause degradation of organic polymers such as rubber and cellulose, it seems likely that some harmful effects do occur and will eventually be detected.

Control

Control of hydrocarbon emission from transportation involves the same treatment of the exhaust fumes as does removal of carbon monoxide. In addition, loss of hydrocarbons by evaporation from the fuel tank and carburettor is reduced by installation of a collection system which eventually returns them to the fuel induction system. Hydrocarbon emissions from stationary sources are controlled by use of afterburning, adsorption, absorption and condensation. Afterburning completes the oxidation of the hydrocarbons to carbon dioxide and water. For adsorption, a bed of activated carbon is used. This bed is periodically cleaned with steam, from which the hydrocarbons may be recovered following condensation. Where possible, the recovered hydrocarbons may be utilized for other processes. Otherwise, they may be disposed of by oxidation of carbon dioxide and water. Absorption is achieved by passing exhaust fumes through a liquid in which the hydrocarbons will dissolve or become suspended. Again, they may be recovered for subsequent use or they may be eliminated by oxidation.

Particulates

Sources and Sinks

Particulates is a term used to cover all small, solid particles and liquid droplets in the air, except droplets of pure water. An alternative term is aerosols. Other terms may be used to define different types of particulate. Mists consist of suspended liquid droplets ; smoke is very small soot particles produced by burning ; fumes are condensed vapours ; dusts are fine particles produced by breakdown of solids. Particulates may also be divided into those that are viable, e.g. bacteria, fungi, moulds and their spores, and those that are nonviable. Viable particulates pose special problems. These are best considered in depth in a microbiology textbook. Hence, the following discussion will relate only to the nonviable particulates.

Nonviable particulates may be formed by the breakdown of material or by agglomeration. Sea-salt aerosols, by far the greatest single source of atmospheric particulates, are formed by the bursting of seawater air bubbles which releases many tiny droplets of seawater. The water quickly evaporates leaving the sea-salt particles in the atmosphere. Other processes producing particles by breakdown are soil erosion leading to dust storms, volcanic activity and combustion. Particulates released directly into the atmosphere by any of above processes are called primary particulates.

Secondary particulates are those formed in the atmosphere by agglomeration, e.g. by the reactions of gases followed by subsequent solution of the products in water droplets. The production of sulphuric acid from sulphur oxides is an example of this. In total, such secondary particulates slightly exceed in quantity the amount of sea-salt aerosol entering the atmosphere each year. Secondary particulates constitute the main anthropogenic contribution to particulate pollution but amount to only 20% of the yearly natural production of secondary particulates. Similarly, anthropogenic production of primary particulates amounts to less than 8% of the natural production on an annual basis. The main source of anthropogenic particulates is generally thought to be industry, followed by domestic burning of coal and other fuels. However, this view is based on the assumption that soil erosion is a natural process. When one considers that erosion has frequently been due to bad agricultural practice such a view is not tenable and it may well be that, on a global basis, agricultural generated soil dust is the principal anthropogenic particulate pollutant.

Particulates inevitable vary enormously in their chemical composition. Soil dust contains largely calcium, aluminium and silicon compounds. Smoke may contain many organic compounds. Secondary particulates frequently contain ammonium sulphate or ammonium nitrate. Some particulates contain pesticide residues, others contain heavy metals. Thus there are hazards quite distinct from the particulate nature of the carrier.

Particulates range in diameter from 0.2 to 5000 nm, but those of diameter less than 100 nm behave like molecules and generally aggregate. Particles of diameter larger than 1000 nm sediment out of the atmosphere under gravity. Thus, most particles in the atmosphere at any time are of diameter between 100 and 1000 nm. Eventually all atmospheric particles reach the earth's surface. Under dry conditions, this will occur by sedimentation, impaction or diffusion. Impaction refers to particles forcibly deposited under the action of wind. Diffusion refers to the normal random motion of small particles in a gas. As a result of this some collide with and adhere to the earth's surface. However, the deposition of most particulates is associated with rain or snow, i.e. wet precipitation. Particles may from nuclei for water condensation and be carried down in the resultant rain, i.e. rainout. Alternatively falling rain or snow will collect particulates as it passes through the atmosphere, i.e. washout. One consequence of increasing levels of sulphur oxides and nitrogen oxides in the atmosphere has been that wet precipitation has become measurably more acidic, especially in industrialized regions. Increased fish mortality in these areas has been attributed to this. Another result may be increased leaching of soil nutrients causing loss of soil fertility.

Toxic Effects

Little is known of the effects of particulates on plants. These are likely to be variable, since the particulates vary so much in chemical composition. However, one general effect will be the reduction of photosynthesis by prevention of light reaching the leaves and by interference with carbon dioxide uptake.

The most marked effects on animals involve the respiratory system. In humans, particles larger than 5000 nm in diameter do not pass beyond the upper respiratory tract. Particles of between 500 and 5000 nm diameter may reach the bronchioles. However, these are removed by ciliary action to the pharynx where they are mostly eliminated through the gastro-intestinal tract by swallowing. Particles less than 500 nm in diameter may reach the alveoli. Such particles may stay there for periods of up to several years since alveolar membranes have no cilia. If particles do penetrate the respiratory tract, they may exert various harmful effects e.g. slowing the ciliary beat and inhibiting the removal of harmful substances in the mucous flow, and thus cause such illnesses as bronchitis. Furthermore, they may be carrying toxic substances. either as minor or major components. In fact, the range of possible toxicities is almost without limit and made more difficult to asses by the potential for synergistic and antagonistic effects.

Control

Attempts to reduce anthropogenic release of particulates must include improved agricultural practices. However, most attention has been paid to industry and four processes have been developed to lower the industrial contribution. The gravity settling chamber is effective in removing particles of diameter greater than 50,000 nm from gases. The cyclone collector uses centrifugal force to remove particles with diameters 5000–20,0000 nm. Wet scrubbers remove as well as liquids and gases, with variable efficiency, depending on the design. Finally, electrostatic precipitators are often an effective way of removing primary particulates.

Fluorides

Fluoride particles are released into the atmosphere by brick factories, aluminium smelters and phosphate works. These particles settle on surrounding pasture and there are many recorded observations of cattle losing teeth and suffering lameness from bone malformations following ingestion of fluoride during grazing. This type of poisoning has been referred to as fluorosis. Plants may also suffer from fluoride poisoning. Gladioli of the variety 'Snow Princess' are particularly sensitive. The leaves mottle and turn brown on exposure to concentrations of hydrogen fluoride of 0.5 $\mu g\ m^{-3}$ or more. There are reports of damage to maize and citrus plants at concentrations as low as 4 $\mu g\ m^{-3}$ in the atmosphere. Since atmospheric fluoride concentration of 1 $\mu g\ m^{-1}$ can lead to plant fluoride concentrations in excess of 30 ppm, it will be seen that chronic exposure to low levels of fluoride may have harmful effects because of fluoride accumulation. Dairy cattle may be particularly at risk.

Deliberate fluoridation of drinking water by public health authorities, with the intention of preventing tooth decay has been a matter or some controversy. Fluoride, as described above, can be very toxic but, at the time of writing, no harmful effects of fluoride have been conclusively demonstrated at the concentrations present in drinking water. Possible correlations with increased incidence of disease have been suggested, but other explanations of the statistical evidence are possible. However, the question remains as to whether the benefit of reduced tooth decay is sufficient to warrant even the remotest possibility of an increase in other, more serious illnesses.

Asbestos Dust

Asbestos dust is an air pollutant which constitutes a considerably localized hazard of the manmade environment. Asbestos is a fibrous material made up of magnesium silicate. It is used in building and in engine gaskets and brake linings. Its main value lies in its resistance to fire, heat and chemical attack. There are two main forms of asbestos, blue asbestos (crocidolite) and white asbestos (amosite, chrysotile, tremolite). Blue asbestos dust is particularly dangerous but dust from both types of asbestos can have serious effects following inhalation. These effects include lung scarring (asbestosis) and cancer of the bronchii, pleura and peritoneum. Although threshold limiting values for asbestos dust in the atmosphere have been established, it seems likely that no concentration can be regarded as completely safe.

Temperature Inversions and Other Environmental Phenomena Related to Atmospheric Toxicants

Temperature inversions

Temperature inversions are mentioned here because they can prevent dispersal of atmospheric toxicants from the site of emission. Generally toxicants can be dispersed by wind action but mountains surrounding a valley can hinder this process, as can buildings in a city. Dispersion then become dependent on vertical movement of air which, in turn, depends on convection. Air immediately above the earth's surface is warmed by the earth, expands becomes less dense than the cooler air above it and rises, allowing the cooler air to move in to replace it. In this way, air currents are produced which can disperse toxicants. Sometimes weather conditions will place a cold air mass below a warm air mass so that convection ceases for a considerable period of time, even for several days. This is the phenomenon known as temperature inversion. Then toxicants are trapped and may accumulate in the cold air mass, the warm air mass is referred to as the inversion layer. It is usually cloudless and can transmit sunlight readily, causing photochemical reactions amongst the trapped pollutants and, therefore, high smog concentrations.

The ozone layer

Besides the concern about direct effects of atmosphere toxicants. There has also been concern about indirect effects. For instance, there has been concern about the stratospheric ozone layer which screens out more than 99% of solar ultraviolet rays. These ultraviolet rays constitute high energy radiation, which could cause serious harm to living organisms. Suggestions have been made that nitrogen oxides from nuclear explosions or from the jet engines of high flying aircraft, and atomic chlorine from the freons used as propellants in aerosol cans, may cause serious depletion of the ozone layer and hence threaten life.

Fig. 3.2 shows the distribution of atomic and molecular oxygen and of ozone as a function of distance from the earth's surface. In the distant stratosphere oxygen is at a low concentration and the rate of formation of ozone shown in equation 12 is reduced ; at the same time the radiation becomes more intense which enhances the decomposition of ozone according to equation 13. As a result atomic oxygen is here the dominant form of odd oxygen.

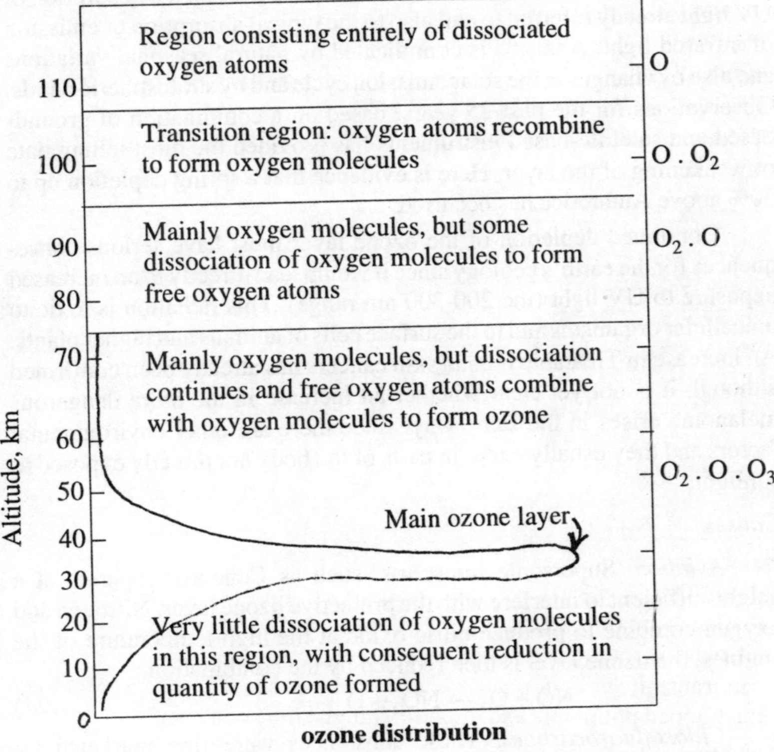

Fig. 3.2. Distribution of ozone and its precursors in the stratosphere

At about 60 km from the earth's surface increasing concentrations of molecular oxygen and nitrogen favour ozone formation (equation 12), while a decrease in the incidence of higher energy light disfavours formation of atomic oxygen. As lower levels are reached ozone becomes the only form of odd oxygen and it forms a layer in the region of 35 km where it accumulates at a concentration of about 1% of the atmosphere. The formation of this layer of ozone extends the absorption of UV light to include the range 200-300 nm and these frequencies are filtered from the sun which falls on the earth. Below the ozone layer in the troposphere the undisturbed system is protected from high frequency light and in its absence the formation of atomic oxygen and hence of ozone is minimal

Depletion of the ozone layer

Measurement of ozone concentrations depends on the absorption of UV light already referred to and also to the typical absorption or emission of infrared light. Analysis is complicated by natural seasonal variations and also by changes in the solar emission cycle and by stratospheric winds. Observations for the past 25 years, based on a combination of ground-based and satellite-based instruments has provided the most telling data of weakening of the layer. Here is evidence that a spring depletion up to 30% above Antarctica has occurred.

Continued depletion of the ozone layer must have serious consequences for the earth's ecology since it would lead directly to an increased exposure to UV light (the 200-300 nm range). This radiation is toxic to unicellular organisms and to the surface cells of animals and higher plants. An increase in Treatable human skin cancers has already been confirmed although it is not yet clear whether an increase in the more dangerous melanoma arises in the same way, since there are other environmental factors and they usually occur in parts of the body not directly exposed to sunlight.

Causes

Aviation. Supersonic transports, such as Concorde, operate at a height sufficient to interfere with the protective ozone layer. Nitrogen and oxygen combine to produce nitric oxide at the high temperature of the engines, the ozone level is then reduced by the combination :

$$NO + O_3 \rightarrow NO_2 + O_2 \qquad (24)$$

Chlorofluorocarbons. These substances were first marketed by General Motors Laboratories in 1930 as they were non-toxic, non-flammable and could replace SO_2, NH_3 and carbon tetrachloride in refrigerators. On the evidence available at the time they appeared ideally friendly to the environment since they were stable to all the reactions which take place in surface air pollution and are unaffected by normal daylight radiation. They came into increasing use as commerical refrigeration became widespread and also found applications as industrial solvents and blowing agents.

Nomenclature. The system is based on the form CFC-*XYZ* where X = number of carbon atoms minus 1 (omitted if $X = 0$) ; Y = number of hydrogen atoms plus 1 ; Z = number of fluorine atoms. SO that $CFCl_3$ become CFC-11 and $CF_2 Cl_2$ is CFC–12.

The two substances of significance are trichlorofluoromethane or CFC-11 and dichlorodifluoromethane or CFC-12. They tend to persist in the troposphere on account of their chemical stability and because they are volatile by choice, insoluble in water and are not removed by rainout. It is now realized that these properties make them undesirable as 'greenhouse' gases but even more seriously it was shown by Molina and Rowland in 1974 that they diffused upwards into the stratosphere and were there exposed to high frequency radiation which caused them to react and destroy ozone (see equation 25-27).

In contrast to their stability in the troposphere the higher energy radiation at 20-40 km causes photolysis :

$$CFCl_3 \xrightarrow{hv} CFCl_2 + Cl^{\cdot} \tag{25}$$

$$CF_2\,Cl_2 \xrightarrow{hv} CF_2\,Cl + Cl^{\cdot} \tag{26}$$

Chlorine atoms produced in this way act to destroy ozone in a chain reaction in an analogous manner to the oxides of nitrogen :

$$Cl^{\cdot} + O_3 \rightarrow ClO^{\cdot} + O_2 \tag{27}$$

As it stands this reaction would have little effect but in the ice clouds which form over Antarctica, chlorine monoxide radicals disproportionate to yield molecular oxygen and chlorine atoms, which re-enter the cycle and can account for very many ozone molecules.

Control Measures. Legislation was enacted in the USA to restrict the use of CFCs as early as 1975 and worldwide production declined through the 1980s. In 1987 an International Protocol was agreed at Montreal whereby the signatories agreed to halve their current production by 1999. In recent years the public and governments have become increasingly concerned. An international work group reported 18% depletion of the ozone layer over Europe in the winter of 1991/2 and work in the USA has confirmed this trend. An increase in skin cancer and eye cataracts has been predicted.

An international meeting at Copenhagen in November 1992 took account of he growing problem and agreed for the most part of phase out CFCs altogether by 1995.

The 'greenhouse effect' of carbon dioxide

It has been suggested that increased concentrations of carbon dioxide in the air, resulting from human activity, may increase the surface temperature of the earth by reducing heat loss. This follows because carbon dioxide absorbs long wavelength infrared radiation emitted by the earth's surface and, in this way, the temperature of the atmosphere is increased. However, while atmospheric carbon dioxide has been increasing, the temperature of the earth has decreased since 1940. This may either be due to increased cloud cover or increased atmospheric particulate concentrations, both of which would reduce the amount of radiation reaching earth. Nevertheless, the potential danger of the 'greenhouse effect' should not be ignored, as it is possible that reduction in the levels of the toxicants discussed in this chapter may leave it as the final atmospheric aberration to be corrected.

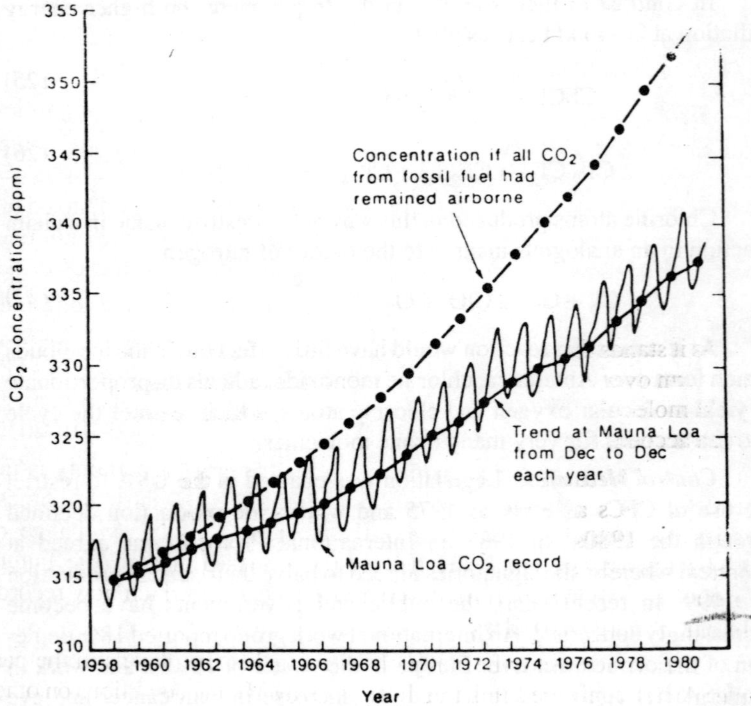

Fig. 3.3. Rise in CO_2 concentration in air at Mauna Loa

The observed increase in CO_2 levels. The natural level of 0.03% corresponds to 275 ppm and correlates with that found in ice cores pre-dating the Industrial Revolution and also with a summary of earlier data made by Callender in 1958. The subsequent upward trend (Fig. 3.3) has been established remote from local discharges at the Mauna Loa observatory following an initiative by Keeling.

These observations are supported by others made at the South Pole and in Alaska at Point Barrow, also by samples taken during aircraft flights.

From Figure 3.3 it is clear that there is a seasonal downward 'wobble' every year due to the uptake of CO_2 during photosynthesis by deciduous trees and plants. It is also apparent that only part of the released CO_2 remains in the atmosphere. This is to be expected since it is soluble in water to the extent of 3.3 g/1 at 0° C and 1.7 g/1 at 20° C. Ocean sediments arise from its solution in sea water but the surface contact with natural water is limited and only moderates the rise in concentration.

Ice cores. The solubility of CO_2 in water ensures that equilibrium is established through time between the levels in air and water. Hence from a knowledge of the level in polar ice of known date it is possible to evaluate the level in air in equilibrium with it at the same period. It has been shown that in the years preceding the Mauna Loa analysis that CO_2 ranged from an unperturbed level of 270 ppm to 320 ppm.

Wood is an important energy source in Third World communities. For example in India a village will consume about 3800×10^9 joules annually and of this almost 90% comes from firewood. Taking India as a whole, wood and charcoal account for one-third of its energy demand but this is small on a global scale since the inhabitants consume only one-twentieth of that consumed by inhabitants of developed countries.

As heavily populated Third World countries such as India and China strive to raise their living standard, residual CO_2 levels must rise at a greater rate than before. At present they account for 80% of the world's population and consume only 25% of commercial energy. Their demand is expected to double during 1990-2010 and redouble during 2010-2030.

Effect of green house warming. A fundamental uncertainty is the past effect of variations in the emissions from the sun whose brightness varies by up to 0.1% in phase with magnetic activity. A change of 0.2-0.5% would account for a fall of 0.4-0.6° C in the Little Ice Age. Future forecasters will benefit from observations by NASA which now records the sun's energy to 0.01%.

It is known from the study of ice cores that CO_2 levels in the past interglacial age were similar to those of today and the conditions on other planets correlate with their surface temperatures. Mars has a thin atmosphere composed mainly of CO_2, a surface pressure less than one-hundredth that of earth and a surface temperature of 223 K. Venus has an atmosphere which is 96% CO_2 at a surface pressure one hundred times that of earth and a surface temperature of 732 K.

The earth's surface temperature has risen by 0.5° C over the last century and predictions depend on forecasting the likely rate of change in the concentration of greenhouse gases. The earth's atmosphere contains 2.6×10^{15} kg of CO_2 to which man added 5×10^{12} kg in 1985 and the current levels is 1.9×10^{13} kg/year. Fossil fuel usage over the past century increased at 4.5%/year, a doubling time of 16 years, leading to the present level of 350 ppm. If the future rate of usage averages 4% then the natural level will be doubled to one of 600 ppm by the year 2030. With growth nearer to the present rate of 2% this point will not be reached until 2050. These predicted dates will vary somewhat depending on the fraction of emitted CO_2 which persists in the atmosphere : at present this is almost one-half.

Computer models which broadly match conditions in the past indicate a temperature rise of 3° C on doubling present CO_2 levels. This must be qualified as there are both positive feedback mechanisms which enhance the effect and negative feedback which reduces it. One example of negative feedback is the reduction of incident radiation as a result of increased cloud cover as oceans warm up. An example of positive feedback in the reduced reflectivity (albedo) owing to the shrinkage of polar ice. Warming is likely to be higher in the northern hemisphere with its higher level of industrial activity and smaller ocean area. At the second World Climate Conference at Geneva in 1990 a forecast was made of 2-5° C warming by the end of next century.

Aside from CO_2 emissions it must be remembered that increases are to be expected in the concentration of other airborne pollutants.

The effect on sea levels. In 1985 responsible predictions based on computer models were of about 1 metre rise by 2050 ; these have since been revised downward. In May 1990 the United Nations group on climate change forecast an increase of 6 cm/decade or an average rise of 20 cm by 2030 and 60 cm by 2100.

Levels rise as a result of expansion of the water body as temperature increase and also of its enhancement by melting South polar ice—melting of North polar ice will not directly change sea level as it is already afloat (a full glass of an iced drink does not spill over if left at room temperature). The North Pole is more vulnerable than the South, which was formed 10 million years earlier, and Arctic ice has been reduced in thickness from about 7 m in 1976 to 4m in 1987.

The Antarctic ice has contributed to sea levels as average temperature have increased. There was an abnormally high release of icebergs in 1930 and another similar period is now underway. The British Antarctic Survey assessed the 24th iceberg to be recorded since the mid-1980s ; it is moving at 3 miles/day and is of exceptional size at 50 miles in diameter and weighing 10^3 billion tons.

The consequences of melting of ice will not be spread uniformly and the predicted average rise of 20 cm will be seen as one of about 35 cm in Europe. This is due to the more rapid disappearance of Arctic ice and its replacement by a body of darker heat-absorbing water. There exist possibilities of drastic changes as a result of the diversion of ocean currents. Warm Gulf Stream waters could take another course to the detriment of the climate of the British Isles and Europe. The direct threat will be to communities on low-lying islands such as the Maldives, 3.66 m above the level of the Indian ocean. In the South Pacific Tonga, Kiribati and Tuvalu are at risk and observation posts have been established there by the Australian Government.

The backing up rivers will lead to flooding. The most serious risk is to Bangladesh in the delta of the Brahmaputra. In England the threat is to the fenlands and the estuaries of the Thames, Humber and Severn. In Europe those at risk include the Garonne, Loire, Seine, Rhine and the Elbe.

The Effect on Agriculture. Changes in the Arctic referred to above will lead to a progressive shift in the meteorological equator at present at 6° N to within 10-12° N. Although this change may be long delayed the consequences are so serious as to demand our attention.

In the UK the south east will become drier and growth of arable crops would tend towards the west. It is also likely that industry would move to the north-west to secure water supplies. Typical French crops such as maize and sunflowers would grow in the south and the malarial mosquito could breed in these areas of England.

The African desert would move northward and areas of Spain, Italy and Greece could become deserts. Similarly, with reduction of winter rainfall, California would come to resemble Mexican desert and the change would also be seen in the Punjab of India.

REFERENCES

Bartell, L.S., and Ritz, C.L., 1974. Stratospheric ozone destruction by man-made chlorofluoromethanes, *Science* 185, 1163-1164.

Bowman, K.P. 1988. Global trends in total ozone, *Science*, 239, 48-50.

Goody, R.M. and Walker, J.C.G., 1972. *Atmospheres*, Prentice Hall, New Jersey.

Hammond, A.L. 1975. Ozone destruction : Problem's scope grows, its urgency recedes, *Science*, 187, 1188-1183.

Jone, Robin Russel and Tom M.L. Wigley (eds.) 1989. *Ozone depletion : Health and Environment Consequences*, John Wiley & Sons, England.

McEwan, M.J. and Phillips, L.F. 1975. *Chemistry of the Atmosphere.* Edward Arnold, London :

Molina, M.J. and Rowland, F.S. 1974. Stratospheric sink for chlorofluoromethanes : chlorine atomic atalysed destruction of ozone, *Nature*, 249 ; 810-812.

Rodhe, H., et, al. 1982. *Tropospheic Chemistry and Air Pollution*, TN 176, World Meteorological Organisation. Geneva.

Tyler Miller, G. (1990) Living in the Environment (6th edition), Wadsworth, London.

Flavin, C. (1989) *Slowing global warming, a worldwide strategy*, Worldwatch Paper, 91, Washington DC.

Grainger, A. (1990) *The Threatening Desert*, Earthscan.

Meetham, A.R. (1981) *Atmospheric Pollution* (4th edition), Pergamon, Oxford (appendix on units).

Warrick, R.A., Barrow, E.M. and Wigley, T.M.L. (1990) *The Greenhouse Effect and its Implications for the European Community*, EC, Luxembourg.

Brimblecombe, P. (1987) *The Big Smoke*, Methuen, London.

CHAPTER 4

Toxic Metals

Introduction

'Toxic metals' is a general collective term usually applied to the elements such as Cd, Cr, Cu, Hg, Ni, Pb and Zn which are commonly associated with pollution and toxicity problems. An alternative (and theoretically more acceptable) name for this group of elements is 'trace metals' but it is not as widely used. Unlike most organic pollutants, such as organohalides, metals occur naturally in rock-forming and ore minerals and so there is a range of normal back-ground concentrations of these elements in soils, sediments, waters and living organisms. Pollution gives rise to anomalously high concentrations of the metals relative to the normal background levels ; therefore, presence of the metal is insufficient evidence of pollution, the relative concentration is all important.

Apart from aerosols in the atmosphere and direct effluent discharges into waters, the concentrations of metals available to terrestrial, aquatic and marine organisms (i.e. their bioavailability) is determined by the solubilization and release of metals from rock-forming minerals and the adsorption and precipitation reactions which occur in soils and sediments. The extent to which metals are adsorbed depends on the properties of the metal concerned (valency, radius, degree of hydration and coordination with oxygen), the physico-chemical environment (pH and redox status), the nature of the adsorbent (permanent and pH-dependent charge, complex-forming ligands), other metals present and their concentrations, and the presence of soluble ligands in the surrounding fluids.

Metals are used widely in electronics, machines and the artefacts of everyday life as well as 'high-tech' applications. Consequently they tend to reach the environment from a vast array of anthropogenic sources as well as natural geochemical processes. Some of the oldest cases of environmental pollution in the world are due to heavy metal use such as

Cu, Hg and Pb mining, smelting and utilization by ancient civilizations, such as the Romans and the Phoenicians.

Biochemical Properties

Some elements are required by most living organisms in small but critical concentrations for normally healthy growth (referred to as 'micronutrients' or 'essential trace elements') but excess concentrations cause toxicity. Those metals which are essential, whose deficiency causes disease under normal living conditions include Cu, Mn, Fe, and Zn for both plants and animals, Co, Cr, Se and I for animals and B, Mo for plants. Most of the micronutrients owe their essentiality to being constituents of enzymes and other important proteins involved in key metabolic pathways. Hence, a deficient supply of the micronutrient will result in a shortage of the enzyme which leads to metabolic dysfunction causing disease.

Some other elements have been shown to have some beneficial effect under rigorous experimental conditions but are not likely to be responsible for deficiency disorder under normal conditions. Elements with no known essential biochemical function are called 'non-essential elements' also known as toxic elements. These elements, which include As, Cd, Hg, Pb, Pu, Sb, Tl and U, cause toxicity at concentrations which exceed the tolerance of the organism but do not cause deficiency disorders at low concentrations like micronutrients. These are clearly shown by the typical dose response curves in Fig. 4.1.

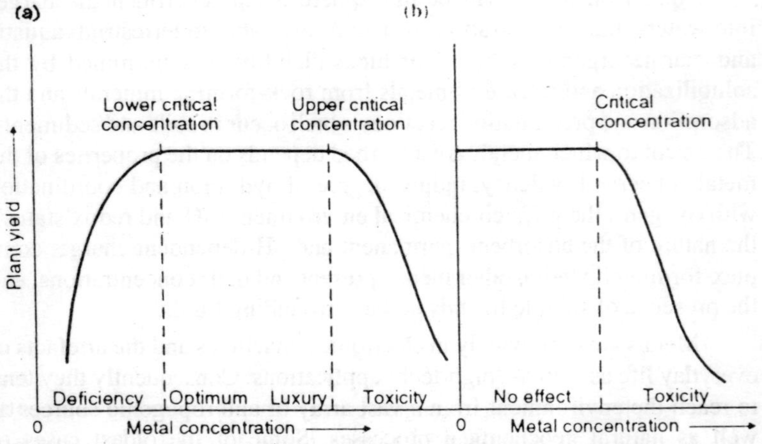

Fig. 4.1 Typical dose-response curve for (a) micronutrients and (b) non-essential trace elements.

At the biochemical level, the toxic effects caused by excess concentrations of these metals include competition for site with essential metabolites, replacement of essential ions, reactions with —SH groups, damage to cell membranes, and reactions with the phosphate groups of ADP and ATP. Organisms have homeostatic mechanisms which enable them to tolerate small fluctuations in the supply of most elements but prolonged excesses eventually exceed the capacity of the homeostatic system to cope and toxicity occurs, which if severe can cause the death of organisms. An example of homeostasis in animals and the control of excess metals is the formation of metallothionein proteins containing —SH groups which bind certain metals, such as Cd and Zn and enable them to be excreted without causing biochemical dysfunction. In plants, similar compounds called phytochelatins carry out the same function binding divalent metals, such as Cd, in physiologically inactive forms.

Sources of Metals

Geochemical Sources

Trace elements together constitute less than 1% of the rocks in the earth's crust ; the macroelements (O, Si, Al, Fe, Ca, Na, K, Mg, Ti, H, P and S) comprise 99% of the earth's crust. The trace elements occur as 'impurities' isomorphously substituted for various macroelement constituents of the crystal lattice of many primary minerals (those found in igneous rocks which originally crystallized from molten magma) In sedimentary rocks trace elements occur sorbed to the secondary minerals which are the products of the weathering (physical disintegration and chemical decomposition) of primary minerals.

Metals present in the atmosphere and the hydrosphere. The concentration of metals in the atmosphere in an area remote from anthropogenic effects (South Pole) are markedly lower than those from various locations in Europe, with its range of industrial and urban pollution, as well as relatively remote rural areas. The maximum concentrations found in the US are quoted for some elements to indicate the highest concentrations monitored for technologically advanced countries. In the case of the chalcophyllic elements (those normally found as sulphides in ore bodies, such as Cd, Cu, Pb and Zn) the concentrations were particularly high near volcanoes (Etna and Hawaii) as a result of the lavas containing high concentrations of these elements.

Anthropogenic Sources

The following are significant sources of metals to the environments.

Metalliferous mining. The metals utilized in manufacturing are obtained from either the mining of ore bodies in the rocks of the earth's crust, or the recycling of scrap metal originally derived from geological sources. Ore bodies are naturally occurring concentrations of minerals with a sufficiently high concentration of metals to render them economically worthwhile exploiting. With increasing demand the environmental impact of the mining operations becomes greater. There will be a need to dispose of greater amounts of tailing (tailings are the finely milled fragments of rock and some ore particles left behind after the extraction of the metal ore concentrates by various methods). Tailings disposal from current mining operations and continued weathering (chemical alteration) of ore minerals in historical and abandoned mining sites is an important source of metals into the environment.

Many of the major metalliferous ores are sulphide minerals and this has several environmental implications. Fragments of these minerals in tailings deposits oxidize on weathering to create acidic solutions which tend to decrease adsorption and hence increase the mobility of the metals in soils, sediments and waters. The oxidation of pyrite (FeS_2) is shown in detail below :

$$FeS_2 + 7/2O_2 + H_2O \rightarrow Fe^{2+}\ 2SO_4^{2-} + 2H^+ \quad (1)$$

$$Fe^{2+} + 5/2H_2O + 1/4O_2 \rightarrow Fe(OH)_3\ (\text{solid}) + 2H^+ \quad (2)$$

$$Fe^{2+} + 1/2O_2 + H^+ \rightarrow Fe^{3+} + 1/2H_2O \quad (3)$$

$$FeS_2 + 14Fe^{3+} + 8H_2O \rightarrow 15Fe^{2+} + 2SO_4^{2-} + 16H^+ \quad (4)$$

The reactions for Zn and Cu ores can be written in a simplified form as :

$$\text{Sphalerite : } ZnS + 2O_2 \rightarrow Zn^{2+} + SO_4^{2-} \quad (5)$$

$$\text{Chalcopyrite } CuFeS_2 + 4O_2 \rightarrow Cu^{2+} + Fe^{2+} + SO_4^{2-} \quad (6)$$

Frequently, bacteria such as *Thiobacillus thiooxidans* help to catalyse the oxidation reactions in tailings deposits and in weathering orebodies.

When the sulphide ore minerals are smelted to obtain the metals, it is necessary to roast the ores in air in order the convert the sulphides to oxides which are then reduced to the metal. This process gives rise to large amounts of SO_2 and metal smelters are among the major sources of this important atmospheric pollutant. In the past, the acidic fumes together with metal aerosol emissions exacerbated the environmental impact of the metals for many kilometres from the smelter source.

Another feature of mining operations is that most of the major ore minerals often have several other metals associated with them. Many of these metals will have contaminated the environment in the vicinity of mines and smelters, although nowadays it is more likely that the minor constituents will also be refined and marketed as well as the major metal constituent of the ores. Old mine waste deposits have been reworked in many countries. A consequence of the reuse of mined material anywhere in the world is the production of large quantities of tailings which need to be disposed of in an environmental appropriate manner. Modern mineral separation methods involve the use of large volumes of water but much of this is normally recycled within the process, although smaller volumes of effluents containing metals, frothing agents and other chemicals (including cyanides in gold extraction) do need disposal eventually.

(*b*) *Agricultural Materials.* Agriculture constitutes one of the very important non-point sources (NPS) of metal pollutants. The main sources are :

- impurities in fertilizers : Cd, Cr, Mo, Pb, U, V, Zn (e.g. Cd and U in phosphatic fertilizers) ;
- pesticides : Cu, As, He, Pb, Mn, Zn (e.g. Cu, Zn and Mn-based fungicides, Hg seed dressings, historical Pb—As orchard sprays) ;
- wood preservatives : As, Cu ;
- wastes from intensive pig and poulty production : Cu, As ;
- composts and manures : Cd, Cu, Ni, Pb, Zn, As ;
- sewage sludge : especially Cd, Ni, Cu, Pb, Zn (but many other elements) ;
- corrosion of metal objects (e.g. galvanized metal roofs and wire fences : Zn, Cd).

Fossil Fuel Combustion. A wide range of heavy metals is found in fossil fuels which are either emitted into the environment as particles during combustion, or accumulate in ash which may itself be transported and contaminate soils or waters, or may be leached *in situ.* Some of the metals arising as pollutants from fossil fuel combustion are Pb, Cd, Zn, As, Sb, Se, Ba, Cu, Mn, and V. The combustion of petrol (gasoline) containing Pb additives gives rise to large amounts of Pb particulates, mainly PbBrCl. Lead-containing particles in the exhausts of petrol vehicles are 0.01 – 0.1 μm in diameter, but these primary particles can cluster to form larger particles (0.3 – 1 μm). Diesel smoke normally contains these larger size particles. The lead-containing particles in motor vehicle exhausts tend to be larger in rural areas and near motorways than in urban areas (Fergusson, 1990).

Coal combustion gives rise to a wide range of metals, including U in some coal which accounts for coal-burning power stations being responsible for the emission of significant amounts of radioactive pollutants. Coal ash can contain relatively high concentrations of soluble compounds, including oxides of B, As, Se which can be leached and cause toxicity in sensitive crops and aquatic organisms. Coal ash is an important sources of Cr (< 172ppm). Crude oil can contain relatively large amounts of V which is emitted during the combustion of some oil products and accumulates in the ash of oil-fired boilers.

Metallurgical Industries. Many metals are used in specialist alloys and steels : V, Mn, Pb, W, Mo, Cr, Co, Ni, Cu, Zn, Sn, Si, Ti, Te, Ir, Ge, Tl, Sb, In, Cd, Be, Bi, Li, As, Ag, Sb, Pr, Os, Nb, Nd and Gd. Hence both the manufacture and disposal, or recycling, of these alloys in scrap metal can lead to environmental pollution of a wide range of metals. Steel manufacture usually involves a lot of recycling of scrap and so steel works are often discrete point sources of atmospheric aerosols of metals.

Non-ferrous metal production causes marked environmental pollution not only of the metals being manufactured, but also of other minor associated metals such as As, Cd, Cr, Cu, Co, Ni, Pb, Sb, Tl, Te, U, V, Zn and Se. Nowadays, many of these metals are extracted from the ores, refined and sold.

Cu smelters have a long history of causing As pollution in the surrounding countryside. In Austria, cancerous conditions have been reported in livestock grazing near to Cu smelters. Zinc ores often contain relatively high concentrations of Cd (<5%) and so Zn production (and, to a lesser extent, Pb, and Cu smelting) can give rise to significant environmental pollution with Cd. Anomalously high Cd concentrations were found in soils and vegetation up to 40 km downwind from historic smelting activities in the Lower Swansea Valley in South Wales and 15 km from the Avonmouth Pb—Zn Smelter near Bristol, in the UK.

Electronics. A large number of trace elements are used in the manufacture of semi-conductors and other electrical components. These include Cu, Zn, Au, Ag, Pb, Sn, Y, W, Cr, Se, Sm, Ir, In, Ga, Ge, Re, Sn, Tb, Co, Mo, Hg, Sb, As and Gd. Environmental pollution can occur from the manufacture of the components and their disposal in waste. There is now a growing industry concerned with the recovery of valuable metals from decommissioned items of complex electrical equipment, such as computers. However, old electronic equipment will include capacitors and transformers containing PCBs, which are highly toxic and persistent organic environmental pollutants.

Other Sources. Other significant sources of heavy metal pollution in manufacture (sometimes in use) and disposal include :

Batteries — Pb, Sb, Zn, Cd, Ni, Hg, Pm

Pigments and paints — Pb, Cr, As, Sb, Se, Mo, Cd, Ba, Zn, Co, I, Ti

Catalysts — Pt, Sm, Sb, Ru, Co, Rh, Re, Pd, Os, Ni, Mo, I
Polymer stabilizers — Cd, Zn, Sn, Pb (from incineration of plastics)
Printing and graphics — Se (xerox process), Pb, Cd, Zn, Cr, Ba
Medical uses : dental alloy — Ag, Sn, Hg, Cu and Zn
drugs/medicinal preparations—As, Bi, Sb, Se, Ba, Ta, Li, Pt
Additives in fuels and lubricants — Se, Te, Pb, Mo, Li

Waste disposal. Many metals, especially Cd, Cu, Pb, Sn and Zn, are dispersed into the environment in leachates from landfills, which pollute soils and groundwaters, and in fumes from incinerators. Sewage sludge contains many metals, including Zn, Cu, Pb, Cr, As and Mo, but the greatest cause for concern is currently considered to be Cd. Although present in many sludges in quite low concentrations (<10ppm), Cd is relatively easily taken up by food crops, especially leafy vegetables, and enters the human diet.

Behaviour of Metals in the Environment

Metals as Atmospheric Aerosol Particles

These remain suspended for varying lengths of time determined by the size of the particle, the windspeed, relative humidity and precipitation. Acrosol particles range in diameter from 5 nm to 20 μm but most are in the size range 0.1—10 μm. Particles > 10 μm tend to settle out under gravity relatively rapidly but those < 10 remain in the atmosphere for 10—30 days being removed by washout, settlement, impaction and, in the case of very small particles (< 0.3 μm) by diffusive deposition. Under some circumstances, such as high humidity, smaller particles may sometimes cluster and form larger particles which are deposited more rapidly. In the 10—30 day period during which aerosol particles may remain suspended in the atmosphere they can be transported thousands of kilometres, depending on the circulation of air masses.

While suspended in the air, these metal aerosol particles may be inhaled by humans and animals and subsequently absorbed into the bloodstream through the alveoli of the lungs. Particles falling onto foliage may also enter plant tissues by absorption through the cuticle but this depends on the presence of moisture and its pH, the type of plant and other parameters. For example, children are at severe risk from the inhalation of aerosol sized particles of lead compounds in the exhaust emissions of motor cars. These particles also settle onto crop foliage and may be consumed with the plant (e.g. on lettuce leaves). Most particles reach the soil eventually and may be ingested with incompletely washed vegetables, by children eating soil intentionally (pica), or accidentally from unwashed adult or children's hands after gardening of playing with contaminated soil.

On being deposited onto the soil, the metal compounds in the aerosol particles react with the soil constituents and become incorporated into the soil system. Metal particles absorbed by plants also reach the soil through the mineralization f plant litter.

Metals in Aqueous and Marine Environments.

Aerosol particles deposited into water, either directly or washed off surfaces into water courses, either react with the constituents of the water or settle to the bottom where they react with the sediments. The solubility of metal ions in solution will depend on the concentrations of anions and chelating ligands, present in the water, its pH and redox status, and the presence of absorbent sediments. Several metal ions are adsorbed and coprecipitated with hydrous oxides of Fe, Mn and Al, in both sediments and soils, for example Fe oxides coprecipitated V, Nn, Ni, Cu, Zn, Mo ; Mn oxides coprecipitated Fe, Co, Ni, Zn, Pb.

Calcium carbonate, either originating from limestone rock fragments in soils and sediments, or precipitation from soil water in the soils of semi-arid and arid regions, also absorbs a range of metals, including V, Mn, Fe, Co and Cd. Clay minerals in soils and sediments are responsible for the adsorption and coprecipitation of V, Ni, Co, Cr, Zn, Cu, Pb, Ti, Mn and Fe.

Metal ions in solution can be absorbed into aquatic plants and animals and can cause toxicity if the concentration is sufficiently high. This factor is exploited in the use of $CuSO_4$. $5H_2O$ to control algal blooms in lakes and reservoirs.

A survey of drinking water in the UK showed that first draw drinking water in 24 towns had Pb > 50 ppm. The worst situation was in Scotland where 34.4% of homes had Pb concentrations in water above the limit (Fergusson, 1990). This is due to the use of Pb pipes and solder and to the water being relatively acidic. The pH of domestic water from public supplies should be between 6.5 and 8.5. The metal content of water in houses is dependent upon the following factors :

- Original metal content, pH, alkalinity and oxygen contents, total hardness and temperature of the water ;
- duration of time the water remains in the pipe (still-stand) ;
- length of flushing time after still-stand.
- materials, age, internal diameter and total length of the pipe in the house, and the type of solder.

Table 4.1. Maximum acceptable concentrations for metal and other inorganic pollutants in drinking waters (WHO, 1984).

	EC/WHO
As	50(W)
B	1000
Ba	100
Fe	50
Mn	20
Cd	5(W)
Cr	50(W)
CN	100(W)
Pb	50(W)
Hg	1(W)
Cu	<3000 (100 at work)
U	—
Zn	<5000 (100 at works)
F	1500 (8—10°C) 700 (25—30°C)

EC = 80/778/EEC Quality of water for human consumption

• (*W*) = *WHO}* 1984 guideline values

Metal ions in soils. Metal pollution can affect all environments but its effects are most long lasting in soils due to the relatively strong adsorption of many metals onto the humic and clay colloids in soils. The duration of contamination may be for hundreds and thousands of years in many cases (e.g. first half lives : Cd, 15—1100 years, Cu 310—1500 years and Pb 740—5900 years depending on the soil type and their physicochemical paramters). Unlike organic pollutants, which will ultimately be decomposed, metals will remain as metal atoms although there speciation may change with time as the organic molecules binding them decompose or soil conditions change.

The extent to which metal ions are adsorbed by cation exchange and non-specific adsorption depends on the properties of the metal concerned (valency, radius, degree of hydration and coordination with oxygen), pH, redox conditions, the nature of the adsorbent (permanent and pH-dependent charge, complex-forming ligands), the concentrations and properties of other metals present, and the presence of soluble ligands in the surrounding fluids.

The selectivity of clay mineral and hydrous oxide adsorbents in soils and sediments for divalent metals generally follows the order Pb > Cu > Zn > Ni > Cd, but some differences occur between minerals and with varying pH conditions. The selectivity order for peat has been shown to be Pb > Cu Cd = Zn > Ca. However, in general, Pb and Cu tend to be absorbed most strongly and Zn and Cd are usually held more weakly, which implies that these latter metals are likely to be more labile and bioavailable.

In general, nearly all metals (except Mo) are most soluble and bioavailable at low pHs and therefore toxicity problems are likely to be more serve in acid environments. In the case of pollution by particles of sulphide ore minerals, the weathering of the sulphide exacerbates the problem by increasing the acidity of the soil. In agricultural soils this situation can be mitigated to a considerable extent by liming.

Methylation of metals in the environment. Arsenic, Hg, Co, Se, Te, Pb and Tl can be methylated in the environment through the action of enzymes secreted by microorganisms (biomethylation) and also by abiotic chemical reactions. There is a possibility that Cd, In, Sb and Bi can also be methylated. The bacteria associated with methylation of these elements are found in the bottom sediments of rivers, lakes and the coastal waters of the sea, soils and the digestive tracts of animals (including humans). This methylation radically affects the behaviour of the elements in the components of the environment, their bioavailability and their toxicology. For example, monomethyl mercury (CH_3Hg^+), the most toxic form of Hg, is lipophyllic and therefore accumulates in body fats and is the only heavy metal to show bioaccumulation along the food chain (Fergusson, 1990).

The uptake of metals by plants. Transfer coefficients (concentration of metal in aerial portion of plant relative to total concentration in the soil) are a convenient way of quantifying the relative differences in bioavailability of metals to plants. However, soil pH, soil organic matter content and plant genotype can have marked effects on metal uptake. The transfer coefficients are based on root uptake of metals but it should be realized that plants can accumulate relatively large amounts of metals by foliar absorption of atmospheric deposits on plant leaves.

Toxic Effects of Metals

The sensitivity of organisms to metal toxicity varies widely with species of plants and animals and genotypes within species (e.g. cultivars of crops) and many factors can modify the response to the toxic dose of metals. Some individuals are genetically adapted to tolerating anomalously high concentrations of certain metals. Homeostatic mechanisms in animals frequently involve special proteins, metallothioneins, which blind with the metals (such as Cd) and render them relatively inactive. Plants have similar compounds called phytochelatins. It is therefore difficult to generalize about toxicity.

Phytotoxicity. The most toxic metals for both higher plants and several microorganisms are Hg, Cu, Ni, Pb, Co, Cd, and possibly Ag, Be and Sn. Although the occurrence of toxicity will depend on soil factors, such as pH, the plant genotype and the conditions under which it is growing (pot in greenhouse, or under field conditions).

The relative toxicity of different elements will be affected by considerable variation in toleration of individuals and, in the case of diets, in the composition of the diets. The doses injected into rats and other experimental mammals probably provide a more accurate comparison of toxicity.

SPECIFIC METALS

Arsenic

Arsenic is a toxic, non-essential element, and occurs widely in nature. It is used in alloys, pesticides, wood preservatives and some medical preparations. It was formerly used in paint pigments, but this use discontinued when it was found that, under damp conditions, moulds converted the arsenic to the highly toxic gases, arsine, AsH_2, and trimethyl arsine, $As(CH_3)_3$.

It is also present in many sulphide ores of metals and is therefore emitted from metal smelters as an atmospheric pollutant (40% of the anthropogenic emissions). Coals also contain significant amounts of As and its combustion accounts for 20% of atmospheric emission. Coal ashes are also a significant source of As which can be leached out into waters or the soil. The toxicological importance of As is partly due to its chemical similarity with P which means that As can disrupt metabolic pathways involving P. Both acute and chronic toxicity are recognized and the continual inhalation of airborne forms of As is known to be carcinogenic. Respiratory cancers have occurred in occupationally exposed workers.

Arsenic is a cumulative poison, causing vomiting and abdominal pains prior to death. It may also cause dermatitis and bronchitis, and may be carcinogenic to tissues of the mouth, oesophagus, larynx and bladder. At the cell level, it can uncouple oxidative phosphorylation and compete with phosphorus in metabolic reactions.

Arsenic is concentrated by organisms exposed to it and accumulates along food chains. Accumulation in fish seems to be favoured by increasing salinity. Anand (1978) has reported arsenic pollution loads in fish sold in Bombay. Crabs and lobsters have been especially noted for accumulating high concentrations, but no cases of human poisoning seem to have arisen from this.

Cadmium

A highly toxic non-essential metal which accumulates in the kidneys of mammals and can cause kidney dysfunction. In humans, kidney damage diagnosed by the presence of microglobulin proteins in the main toxic effect resulting from chronic exposure to the metal. High concentrations of inhaled Cd aerosols can cause emphysema and related acute lung conditions. Cadmium becomes very volatile above 400°C and hence is likely to be dispersed as an aerosol when mixtures of metals containing Cd are heated or cast.

Cadmium is a normal constituent of soil and water at low concentrations. It is usually mined and extracted from zinc ores, especially zinc sulphide. Industrially, cadmium is used as an anti-friction agent, as a rust proofer and in alloys. It is also used in semiconductors, control rods for nuclear reactors, electroplating bases, PVC manufacture and batteries.

It tends to be less strongly adsorbed than many other divalent metals and is therefore more labile in soils and sediments and more bioavailable. There is more danger from this metal moving through the human food chain from contaminated soils than most other metals. Sewage sludge-amended soils can contain sufficiently high concentrations of Cd to cause elevated concentrations of Cd in food crops and there is a European Community Directive limiting the maximum Cd content of sludged soils to 3 ppm. Although sewage sludge applications are considered a major source of Cd in the soils receiving sludges, the most important sources overall are phosphatic fertilizers and industrial emissions.

In the environment, cadmium is dangerous because many plants and animals absorb it efficiently and concentrate it within their tissues. Normally, however, retention from food by mammals is low but absorption is increased if the mammals are on a low calcium diet. Once absorbed, cadmium associates with the low molecular weight protein, metallothionein, and accumulates in the kidneys, liver and reproductive organs. Very small dose can cause vomiting, diarrhoea and colitis. Continuous exposure to cadmium causes hypertension, heart enlargement and premature death. There is some evidence suggesting that cadmium can induce chromosome abnormalities and may exert a carcinogenic effect on the lungs.

A serious case of Cd poisoning occurred in the Jintsu Valley in the Toyama Prefecture in Japan, where Pb-Zn mining had caused widespread Zn and Cd contamination on the alluvial soils, most of which were used for paddy rice production. The farmers in the valley live mainly on rice grown in the contaminated paddies and also relied on the metal-polluted river for their drinking water. After the Second World War, it was found that more than 200 elderly women who had all had several children had developed kidney damage and skeletal deformities. The condition was known as 'itai-itai' disease which literally means 'ouch-ouch' due to the pain caused by the deformed bones. The Cd toxicity was exacerbated by a low protein and vitamin D diet and the birth of several children. The rice which they consumed contained ten times more Cd than local controls and the contaminated water was an additional intake. It was estimated that the people in the valley had a Cd intake of around 600 μg Cd/day which is around ten times greater than the maximum tolerable intake of 60—70 μg/day. A survey of paddy soils in the whole of Japan revealed that 9.5% of the area was significantly contaminated with Cd, with a further 3.2 % of upland rice-growing soils and 7.5% of orchard soils. The source of the Cd in these soils is probably phosphate fertilizers and industrial/mining pollution.

Higher doses of Cd cause kidney problems, anaemia and bone marrow disorders. The major portion of Cd ingested into our body is trapped in the kidneys and eliminated. A small fraction is bound effectively by the body proteins, metallothionein which present in the kidneys, ; the rest is stored in the body and gradually accumulates. It excessive amounts of Cd^{2+} are ingested, it replaces Zn^{2+} at key enzymatic sites, causing metabolic disorders.

This occurrence of cadmium in river waters in different parts of the world is reported to be in the range 1 to 100 mg Cd per kg of the sample. Srivastava and Jaiswal (1989) have reported that whereas cadmium retards growth in the water plant *Spirodelapolyrrhiza* it induces formation of vegetatively propagating organ called *turion*. This is a distinct adaptive strategy to ensure successful recurrence of the species through turions after the stress is over in the water environment. They have noted the cadmium toxicity in this species is in the concentration range 0.005 ppm to 2.00 ppm and the adverse effects are on pigmentation and protein concentration.

The concentration of Cd in sea water on an average is 0.15 mg per litre (Goldberg, 1976). Anand (1978) has reported 16-17 mg Cd per kg of sample in fish sold in Bombay which is of the same order reported in other parts of the world. The toxic levels are 200 mg Cd per litre of water.

Barium

The most common naturally occurring form of barium is barium sulphate, sometimes called barite or baryta. Barium derivatives are used as fillers for rubber, linoleum, etc., as pigments in paint, in glass manufacture, in the ceramic industry and in various other industrial applications. The applications constitute the main source of risk, primarily to workers in the appropriate industries. Barium salts are highly toxic. They cause vomiting and diarrhoea, which may be associated with stomach, intestinal and kidney haemorrhage. They also affect the central nervous system, causing convulsions and have been implicated as a cause of pneumoconiosis. Despite the known toxicity of barium salts, barium sulphate is used to cost the alimentary tract for X-ray photographs since barium absorbs X-rays strongly and thus increases the contrast. The insolubility of barium sulphate minimizes its toxicity, and makes its use acceptable.

Beryllium

Beryllium is released during the burning of coal, but the main environmental hazard is to those working in industries where beryllium is produced or used, e.g., the manufacture of nuclear reactors, aircraft and rockets. In humans, beryllium has been shown to damage skin and mucous membranes. It accumulates in the lungs where it causes beryllium disease (berylliosis).

It may cause cancer in lungs and bone marrow. It is not excreted from mammalian tissue and, therefore, its effects are cumulative. At the biochemical level, beryllium competes with magnesium for enzyme sites and has been shown to inhibit DNA polymerase, thymidine kinase and alkaline phosphatase.

Chromium

Like most of the metals discussed here, chromium is widely distributed. For most organisms, it is essential as a micronutrient in trace quantities for fat and carbohydrate metabolism. In industry, it is used in making steel alloys, in chromium plating and in leather tanning. Chromates are water soluble and can poison sewage treatment processes. The chromium ion can exist in four valency states ; C_1^{2+}, C_1^{3+}, C_1^{5+} and C_1^{6+}. Of these, the hexavalent ion is the most toxic and it should be reduced to the trivalent state to form insoluble products before chromium waste is released into the environment. Soluble chromate concentrations of 0.5 μg/ml have been shown to be significant. Hence the redox conditions in the environment are very important, waterlogged soils with reducing soils will have the less toxic Cr(III). However, in many freely drained aerated soils the predominant form is also Cr(III) because soil organic matter results in the reduction Cr(VI) to Cr(III). Soils developed on ultramafic rocks, such as serpentinites can contain very high concentrations of Cr of geochemical origin and cannot be considered polluted but either 'naturally' or 'geochemically' enriched.

Chromite (III) appears to be more toxic to fish than Cr(VI), especially salmon, but toxic concentrations for several species of fish range from 0.2—5 μg/ml. Municipal wastewater can contain concentrations of < 0.7 μg/Cr ml, mainly in the Cr(VI) form, which are toxic to many species of marine animals, algae and microorganisms but reduction of Cr(VI) to Cr(III) usually occurs if there is organic matter present (Langard, 1980).

Chromium is carcinogenic, causing cancer of the respiratory organs in chromate workers chronically exposed to Cr-containing dusts (Langard, 1980). Hexavalent chromium has been implicated in poisoning in Japan. In this case, aerosols from chromium refining plants appear to have affected a considerable number of people causing lung cancer. Besides this, it has been shown that chromates act as irritants to the eyes, nose and throat, and chronic exposure may lead to liver and kidney damage. A characteristic effect on human beings is the appearance of perforations in the nasal septum. At the cell level it appears that hexavalent chromium may cause chromosome abnormalities. Chromium is particularly dangerous because it accumulates in many organisms. Some aquatic algae have been shown to concentrate it 4000 times above the level of their immediate environment.

The maximum permissible limit of Cr in drinking water as recommended by the W.H.O, is 0.05 mg per litre. Rai and Raizada (1987 and 1988) have studied biochemically the effect of nickel, silver, chromium and lead on toxicity and its regulatory mechanisms on the nitrogen fixing blue geen algae particularly *Nostoc muscorum* when treated singly and in different combinations.

Copper

Copper is one of the most abundant trace metals. It is widely used in its metallic state, either in the pure form or in alloys. For almost all organisms it is an essential micronutrient. It may occur in very high concentrations in water, sediments and biota in some localized areas as a result of mining activities, of intensive use of copper pellets in pig rearing, or of the application of copper fungicides. However, there is no evidence of food chain magnification. Hence, most toxic effects are due to immediate exposure to the element. All organisms are harmed by excessive concentrations, which may be as low as 0.5 ppm for algae. Most fish are killed by a few parts per million (Lopez and Lee, 1977). In higher animals brain damage is a characteristic feature of copper poisoning.

Copper can be deficient in some soils causing severe loss of yield in several crops, especially cereals. Toxicity problems can occur in crops in polluted soils and in livestock grazing herbage growing on polluted soils. sheep are the agricultural livestock most sensitive to Cu toxicity, but they (and cattle) are also prone to deficiency disorders. Herbage with < 5 ppm Cu can lead to Cu deficiency in both sheep and cattle but if the herbage contains > 10 ppm Cu then toxicity is likely to occur in sheep. Copper pollution can arise from Cu mining and smelting, brass manufacture, electroplating and excessive use of Cu-based agrichemicals (e.g. Bordeaux Mixture). Copper sulphate is used widely as an algicide in ornamental ponds and even in water supply reservoirs which are affected by blooms of toxic blue-green algae. Copper is used widely in houses for piping water and although the concentrations of Cu in the Drinking water is higher in soft water, this is not considered to be a hazard so long as the pH is within the normal limits (pH 6.5—8.5). More acid waters could create problems with excessive concentrations, but none have been reported with public water supplies.

Pillai (1985) has reported the presence of copper in Periyar river water, sediments and fish from South India. The upstream region water contained 0.05 to 0.1 ppm copper and the industrial area 0.025 to 0.3 ppm while the sediment of the respective zone had 1 to 100 and 5 to 100 ppm copper. Fish contained about 0.7 ppm. Copper occurs in water in different forms. Humic acid, present in natural waters is known to form complex with copper. Copper complexes are less toxic to water plants and animals in the pH range 6—8. Copper-humic acid form is common in freshwaters while $(Cu(OH)_2$ and $CuCO_3$ are characteristic forms in sea water.

Iron is the fourth most abundant element in the earth's crust. Its greatest use is for structural iron and steel, but it is also used for making dyes and abrasives. It is an essential micronutrient required in trace quantities for the normal metabolism of plants and animals. It is a constituent of cytochromes and nonheme iron proteins involved in photosythesis, n_2 fixation and respiratory linked dehydrogenases (Noggle and Fritz, 1986) Ingestion of excessive amounts may result in the inhibition of activity of many enzymes. The amounts consumed must be very large because only a small proportion of all iron ingested is absorbed from the gastro-intestinal tract. Inhalation of iron dust can cause benign pneumoconiosis and can enhance harmful effects of sulphur-dioxide and various carcinogens.

Many streams are poisoned by high levels of iron in acid mine drainage. Pyrite, iron sulphide, is often found in close association with coal deposits. Upon exposure to moisture and atmospheric oxygen, the ferrous iron is oxidized to the ferric state, a reaction which is frequently accelerated by bacteria of the Thiobacillus—Ferrobacillus group. The ferric iron can then react with sulphide in the presence of water to produce sulphuric acid, or react directly with water to produce a yellow, flocculent mass of ferric hydroxide. Beside being acidic, water affected in this way becomes deficient in oxygen. Such poisoning of streams is reckoned to be one of the main causes of fish kill. Although particularly associated with mining, streams running through iron-laden strata may become poisoned spontaneously.

Lead

A non-essential element ; it is a neurotoxin and a good example of a multimedia pollutant. The main sources of Pb pollution in the environment are petrol (air pollutant, but can also be water or soil pollutant from spillages), particulates in exhaust fumes from petrol combustion (air pollutants, inhaled by humans), particulates from petrol, fossil fuel combustion in soil (soil pollutant — taken up by plants and also ingested with plant food crops), paint flakes from old paint containing a percentage of Pb, Pb in some traditional ethnic cosmetics (e.g. **surma** — skin absorption), constituent of solders and varnishes used on interiors of food can (food contaminant), Pb pipes for potable water (water pollutant), pesticide (historic use of Pb and As containing pesticide sprays in orchards), lead shot used in guns for game and clay pigeon shooting (soil pollutant, but also a food contaminant if inadvertently consumed with the game flesh) and, finally, Pb pollution from mining and smelting of the ore (usually PbS)—this includes acid mine drainage with soluble Pb (water pollutant), tailings from ore dressing (particulate water pollutant and soil pollutant, weather to release soluble Pb), and smelter fumes—Pb aerosols (air and soil pollutants).

On a comparative basis, Pb is neither as toxic as many other metals nor as bioavailable, however, it is generally more ubiquitous in the environment and is a cumulative toxin in the mammalian body, so toxic concentrations can accumulate in the bone marrow, where red blood corpuscle formation (haematopoiesis) occurs. At least five stages in the formation of the haem part of haemoglobin are affected by Pb but the two enzymes most affected are δ-amino laevulinic dehydratase (ALAD) and ferrochetalase (Waldron, 1980). This inhibition of haem synthesis results in anaemia. Kidney damage also occurs as a result of exposure to Pb. Lead, like Hg, is a powerful neurotoxin and a range of pathological conditions are associated with acute Pb poisoning, most characteristic of which is cerebral oedema. However, the absorption of Pb in amounts which are not high enough to cause acute poisoning may induce behavioural abnormalities, including learning difficulties.

The critical concentrations for Pb in blood are the EC recommended level of 35 μg/dl and the UK threshold level for follow-up investigations of 25 μg/dl in at least half the population and less than 30 μg/dl in 90% of the population. The critical level in blood for occupational exposure is higher with values of 60, 80 and even 100 μg/dl being used in different countries.

Mercury

Mercury is concentrated in various ores, the principal one being cinnabar (HgS). It has been mined since 700 BC and is currently used industrially in three forms : as the metal, in organic compounds and in inorganic compounds. The greatest use of mercury is in the production of electrical apparatus. The second greatest use is in the chloro-alkali industry, which produces chlorine and caustic soda by electrolysis of sodium chloride solution using mercury as the cathode of the electrolysis cell. The third greatest use worldwide is in fungicides.

Nearly all the mercury used by man eventually enters the natural environment. To this may be added amounts released during the production of mercury and other metals, from coal burning and from weathering of rocks.

All forms of mercury are potentially toxic but the toxicities vary considerably. The least toxic are the inorganic mercury compounds. They are not readily absorbed from the gastrointestinal tract. Once absorbed they may accumulate in the liver and kidney but normally they are excreted quite rapidly in the urine. It is worth noting that mercury amalgam has long been used in dentistry to fill teeth without any toxic effects being noted.

Hg^{2+} ion, is fairly toxic. It has high affinity for sulphur atoms, and easily attaches itself to the sulphur containing amino acids of proteins. It also forms bonds with haemoglobin and serum albumin, both of which contain sulphydryl groups. This ion, however, does not travel across membranes and hence does not get access to cells.

Mercury vapour is the most hazardous of the inorganic forms because it can diffuse through the lungs into the blood and then into the brain, where serious damage can occur. Arylmercurials are as toxic as the inorganic forms since they are readily broken down to inorganic derivatives in the tissues. Alkylmercurials RHg^+ are the most toxic mercury compounds so far studied. They are fairly stable and have long retention times in the tissues. Therefore, they readily accumulate to high concentrations. Their lipid solubility gives-them an affinity for nervous tissue which accounts for many of their harmful effects. In addition, they have been reported to cause abnormalities in cell division and to increase the frequency of chromosome breakages. Some of these abnormalities may be due to combination of mercurials with sulphydryl groups. Inhibition of enzymes in this way has been demonstrated frequently. So far, no effective treatment for mercury poisoning has been developed, though British antilewisite (BAL) or calcium ethylenediaminetetraacetate (Ca_2EDTA) may have some alleviating effect.

Biological Methylation

Unfortunately, in assessing the risk from mercury in a particular environment, it is not enough to know the form in which it entered that environment because various transformation can take place. Probably the most serious of these is the transformation of metallic mercury to methyl and dimethyl derivatives by anaerobic micro-organisms, especially *Clostridium cochleariul*, in aquatic sediments. This may also occur in decaying fish the transformation is facilitated by Co(III)- containing vitamin B_{12} coenzyme. A CH_3– group bonded to Co(III) on the coenzyme is transferred enzymatically by methyl cobalamin to Hg^{2+}, yielding CH_3Hg^+ or $(CH_3)_2$ Hg. Under aerobic conditions, this transformation can be brought about by *Pseudomonas* spp, and by the fungus, *Neurospora crassa.* In essence, it represents conversion of the least toxic form of mercury to the most toxic. Other transformations which can be brought about by bacteria are the following :

Phenyl, ethyl and methyl mercury can be reduced to elemental mercury and benzene, ethane and methane, respectively ;

Phenylmercuricacetate can be converted aerobically to elemental mercury and diphenyl mercury ;

Mercuric ions can be reduced to elemental mercury. Not all the transformations observed are biological. Under alkaline conditions, methyl mercury converts to the more volatile dimethyl mercury. Under oxidizing conditions, in the presence of ultraviolet light, phenyl mercury, alkoxyalkyl mercury and alkyl mercury may break down to give inorganic mercury. Under anaerobic conditions, mercuric ions may combine with hydrogen sulphide to form poorly-soluble mercuric sulphide. With subsequent aeration, mercuric sulphide can be converted to the soluble sulphate which may then be methylated biologically.

Most macro-organisms are relatively insensitive to mercury and its derivatives. Nitrogen fixing bacteria in the soil require levels of about 100 ppm before they are adversely affected. Normal soil levels are between 0.005 and 1 ppm. However, marine and freshwater phytoplankton, especially diatoms, are very sensitive to organomercurial fungicides and as little as 0.001 ppm may reduce their photosynthetic efficiency. Many algae and other plants have the ability to absorb and concentrate mercury from the surrounding environment. Droplets of elemental mercury have been found in chickweed. Such high concentrations may cause mitotic disturbances and kill the plants. Fortunately, most agricultural plants do not seem to absorb much mercury. Animals tend to accumulate mercury through there food. Pike* can accumulate a concentration of mercury 3000 times higher than that in the water in which they live. Tuna and Swordfish show the same ability. Similar observations have been made on predatory birds. Seed-eating birds accumulate mercury where alkylmercury seed dressing are being used. Much less is accumulated where alkoxylakyl compounds are used, and negligible quantities where inorganic mercury compounds are used.

The effects of mercury and it derivatives on man deserve special consideration because it was mainly these that caused concern about the effects of heavy metals released into the environment in large amounts. The first serious incident to come to light occurred at Minamata Bay in Japan. In this case, comparatively nontoxic inorganic mercury along with some methyl mercury was released in effluent by a chemical factory using mercuric sulphate catalysts in acetaldehyde production. The effluent entered a river running into Minamata Bay. In the sediments, the inorganic mercury was converted to methyl mercury. The accumulated in shellfish and fish which were eaten by the local inhabitant. Consequently by 1975, 116 people had died and many were left paralysed for life. Others suffered impairment of vision and hearing and other neurological symptoms. Prenatal poisoning of the foetus was observed even in the absence of symptoms in the mother. Since the Minamata Bay incident, another has occurred around the Agana River, Niigata, Japan. This led to 23 deaths. In both these cases, many domestic animals, fish, shellfish and seabirds were affected.

Mercury poisoning of human beings has also been caused by the consumption of food containing high concentrations derived from alkylmercury agricultural seed treatments used to prevent seed-borne disease. In Iraq, treated seeds, intended for planting, were used to make bread. Thousands were poisoned and hundreds died. In the USA, cattle fed on treated grain were slaughtered for human consumption. Again, severe poisoning resulted. Consequently a number of countries have now banned the use of alkylmercurial seed treatments. A committee of experts constituted by the Food and Agriculture Oganization (FAO) and World

* *A pike is a large fish that lives in rivers and lakes, and that eats other fish.*

Health Organization (WHO) has recommended that the use of alkylmercurials should be restricted to stocks of cereal seed used for plant breeding or seed production, and never permitted for treatment of cereal seed for export for the production of food.

Despite the major incidents referred to above, it seems fairly certain that the average person is at no great risk from exposure to mercury. The normal dietary intake a well below what is thought to be the tolerable limit of 0.3 mg per person per week, of which not more than 0.2 mg should be in the methylated form, according to WHO. For most people, the chances of appreciable exposure from air, pesticides or pharmaceuticals are very limited. Where there is a risk of occupational exposure, this should be minimized by appropriate precautions and medical screening. With regard to the general environment, there is still a need to know more about the concentration and distribution of mercury, especially with regard to those areas where localized high concentrations do exist.

Manganese

Manganese ocurs widely in nature and is of considerable importance in the manufacture of steel. Biologically, it is an essential micronutrient for most organisms. It is required for activity of some dehydrogenases, decarboxylases, kinases, oxidases, peroxidases etc. It is required for photosynthetic evolution of oxygen. This metal is not a serious pollutant as in most waters its concentrations is quite low ranging from 0.005 to 1 mg per litre. Potassium permanganate is used in very small doses to disinfect well water particularly in Indian villages. Excessive amounts of manganese affects animals adversely, causing cramps, tremors and hallucinations, manganic pneumonia and renal degeneration.

Nickel

Nickel is used in various forms for nickel plating, as a catalyst, as a mordant and in ceramic glazes etc. Again it is a micronutrient for most organisms but excessive quantities have toxic effects. In animals these include dermatitis and respiratory disorders, including lung cancer following inhalation. Amongst enzymes inhibited are cytochrome oxidase, isocitrate dehydrogenase and maleic dehydrogenase. A particularly poisonous derivative of nickel is nickel tetracarbonyl, $Ni(CO)_4$.

Selenium

Strictly speaking selenium is not a metal, though it has certain metallic properties. It is a member of the sulphur-group, produced as a byproduct of the extraction of copper, nickel, gold and silver ores. It is used in electronics, and in paints and rubber compounds, for most organisms it is an essential micronutrient but it can be toxic at very low concentrations. The maximum permissible concentration in drinking water is 0.01 ppm. Poisoning of livestock has occurred where cattle have eaten plants of the Brassica family which have taken in selenium and incorporated it in cysteine and methionine in place of sulphur, Selenium itself irritates the eyes, nose, throat and respiratory tract. It can cause cancer of the liver, pneumonia, liver and kidney degeneration and gastro-intestinal disturbance.

Tin

Tin is widely used mainly for making tinplate and various alloys and compounds. It is an essential micronutrient and the main cause for concern regarding its toxicity has been the development of trialkyl-tin and triaryl-tin compounds having powerful biocidal properties. These are used on growing crops as fungicides and insecticides. They are also used as antimicrobial agents and in marine anti-fouling paints. They can damage crops and, in animals, accumulate in the central nervous system with harmful effects.

Vanadium

Vanadium is widely distributed. It is used as an alloying element for steel and iron, in making oxidation cataysts and in colouring agents used in the ceramic industry. Large amounts enter the atmosphere from the burning of some petroleums. It is an essential micronutrient and may be accumulated by some marine organisms to concentrations many times higher than those in the surrounding water. Excessive levels in animals inhibit tissue oxidation and synthesis of cholesterol, phospholipids and other lipids, and amino acids. Such levels may also cause precipitation of serum proteins.

Zinc

Zinc makes up only 0.004% of the earth's crust. Its most important use is as a protective coating on other metals, particularly in galvanizing iron and steel. It is an essential micronutrient and an essential constituent of alcohol dehydrogenase, carbonic anhydrase, alkaline phosphatase, carboxy peptidase B, and other enzymes (Noggle and Fritz, 1986). In most natural waters zinc is found in traces (less than 1 mg per litre *i.e.*, well within the safe limits). Concentrations above 5 mg per litre causes disagreable taste, In drinking water the level of zinc usually ranges from 0.005 to 1 ppm or mg per litre, but in certain regions it may exceed upto 7.0 mg per litre. Zinc is generally regarded as one of the less hazardous elements, though its toxicity may be enhanced by the presence of arsenic, lead, cadmium and antimony, as impurities. Toxic effects have been observed from the inhalation of fumes from galvanizing baths. The 'zinc fever' produced is characterized by chills, fever and nausea. Removal from the fumes leads to complete recovery. Zinc chloride fumes have sometimes caused fatal oedema of the lungs. Zinc or galvanized containers are not recommended for food storage but are acceptable for storing drinking water. This is because acidic foods can dissolve enough zinc from the container to cause poisoning. A factor which serves to minimize the risk of zinc poisoning is that it appears to be lost along food chains, unlike methyl mercury or cadmium, for example, which accumulate.

REFERENCES

Anand, S.J.S. 1978. Determination of Mercury, Arsenic and Cadmium in fish by neutron activation. *J. Radio, Anal. Chem.* 44, 101-107.

Goldberg, E.D. 1976. *The Wealth of the Oceans*, The UNESCO Press Paris, 172.

Lopez, J.N. and G.F. Lee. 1977. Environmental Chemistry of Copper in Torch Lake, Michigan, *Water, Air* and *Soil Pollution*, 8, 373-385.

Noggle, G.R. and G.J. Fritz 1986, *Introductory Plant Physiology*, Prentice-Hall of India Pvt. Ltd., New Delhi.

Pillai, K.C. 1985. Heavy metals in aquatic environment. In C.K. Varshney (Ed.) 1985. *Water Pollution and Management*, Wiley Eastern Limited, New Delhi.

Rai, L.C. and M. Raizada. 1987. Toxicity of nickel and silver ions to *Nostoc muscorum* : Interaction with ascorbic acid, glutathione and sulphur containing amino acids. *Ecotoxiology and Environmental Safety,* 14, 12-20.

Rai L.C. and M. Raizada. 1988. Impact of Chromium and lead on *Nostoc muscorum :* Regulation of toxicity of ascorbic acid, glutathione and sulphur containing amino acids. *Ecotoxiology and Environmental Safety,* 15, 21-32.

Srivastava, Alka and V.S. Jaiswal. 1989. Effect of Cadmium on turion formation and germination of *Spirodela polyrrhiza L.J. Plant Physiology*, 134, 385-387.

Alloway, B.J. (ed) (1990) *Heavy Metals in Soils*, Blackie and Son.

Bowen, H.J.M. (1979) *The Environmental Chemistry of the Elements*, Academic Press, London.

Fergusson, J.E. (1990) *The Heavy Elements : Chemistry, Environmental Impact and Health Effects*, Pergamon Press, Oxford.

Waldron, H.A. (ed) (1980) *Metals in the Environment*, Academic Press, London.

Bowen, H.J.M. (1979) *The Environmental Chemistry of the Elements*, Academic Press, London.

Fergusson, J.E. (1990) The Heavy Elements : Chemistry, Environmental Impact and Health Effects, Pergamon Press. Oxford.

Harte, J., Holden, C., Schnieder, R. and Shirley, C. (1991) *Toxics A to Z*. University of California Press, Berkeley and Los Angeles.

Kabata-Pendias, A. and Pendias H. (1984) Trace *Elements in Soils and Plants*, CRC Press, Boca Raton, Fl.

Kazantis, G. (1980) Chapter 8 in Waldron, H.A. (ed) *Metals in the Environment*, Academic Press, London.

Kloke, A., Sauerbeck, D.R. and Vetter, H. (1984) in Niragu, J.O. (ed) *Changing Metal Cycle and Human Health*, Springer-Verlag, Berlin.

Krauskopf, K.B. (1967) *Introduction to Geochemistry*, McGraw-Hill, New York.

Langard, S. (1980) Chapter 4 in Waldron H.A. (ed) *Metals in the Environment*, Academic Press, London.

Manahan, S.E. (1991) *Environmental Chemistry* (5th edition), Lewis Publishers, Chelsea, Mich.

O'Neill, P. (1990) Chapter 5 in Alloway, B.J. (ed) *Heavy Metals in Soils*, Blackie and Son, Glasgow.

Waldron, H.A. (1980) Chapter 6 in Waldron. H.A. (ed) *Metals in the Environment*, Academic Press, London.

CHAPTER 5

Radionuclides

Introduction

Radionuclides are elements which spontaneously disintegrate (decay) into smaller particles and emit ionizing radiation. Certain types of this radiation, e.g., alpha particles and neutrons, have a high energy transfer per unit path length. This means that they do not travel far from the source, i.e., no more than 70 μ m in soft tissue, but have pronounced effects on any atom or molecule with which they come in contact. Even at low doses the effects can be very serious, but the risk is highly localized. Beta particles (electrons) can travel farther, i.e., up to a few mm, but are still not very penetrating. Their effects at low doses can also be serious. On the other hand, high energy X-or gamma-rays are very penetrating but, because of their low energy transfer per unit path length, tend to have little effect at low doses, e.g., the 27 to 30 millirads usual in medical chest X-ray. The rad (radiation absorbed dose), is defined as 0.01 J kg^{-1}, i.e., it is a measure of the amount of radiation energy absorbed per unit mass of material. However, one rad of alpha-radiation is about four times as damaging as one rad of X-rays. Therefore, in order to compare doses in terms of biological effects, another unit is used, the rem. One rem is biologically equivalent to one rad of X-rays. As the rad and rem are very large units relative to normally occurring radiation, most of this is measured in millirads or millirems. Recently, SI units for radiation have been introduced. In these, the gray (Gy) replaces the rad, and the sievert (Sv) replaces the rem. The relationship between these units is shown in Table 5.1.

The disintegration rate of a radionuclide is defined in terms of its half life, which is the time taken for half the atoms in a given sample to disintegrate. Radionuclides with short half lives, i.e., up to a few days, may be very dangerous when produced, but the danger does not last. Those with very long half lives, i.e., more than 10^5 years, are usually of such low activity as to be fairly safe. Therefore, those with intermediate half lives have the greatest environmental significance.

Table 5.1 Units for Measurement of Radioactivity

Physical quantity	*SI unit*	*Non-SI unit*	*Relationship*
Activity, i.e., nuclear transformations	becquerel (Bq) $1\ Bq = 1\ s^{-1}$	curie (Ci)	$1\ Bq = 2.7 \times 10^{-14}\ Ci$ $1\ Ci = 3.7 \times 10^{10}\ Bq$
Absorbed dose	gray (Gy) $1\ Gy = 1\ J\ kg^{-1}$	rad	1 Gy = 100 rads 1 rad = 0.01 Gy
Dose equivalent	sievert (Sv) $1\ Sv = 1\ J\ kg^{-1}$	rem	1 Sv = 100 rems 1 rem = 0.01 Sv

Radionuclides occur naturally in the environment. Radonuclides are also produced directly or indirectly from human use of uranium. As a result of mining and processing ores to produce usable radioactive materials, large piles of waste 'uranium tailings' are left behind and radionuclides can be leached from these by rainfall, etc. Atmospheric explosions of nuclear weapons produce fallout of radionuclides over large areas of the earth's surface. Two components of this fallout have caused considerable concern, i.e., strontium-90 and caesium-137. In nuclear power plants radionuclides are formed when impurities in the primary coolant water are bombarded with neutrons from the fuel elements in the core. Even if the coolant water is free of impurities radionuclides may escape into it from the steel or zirconium containers holding the radioactive fuel. Thus, the coolant water must be constantly monitored so that release of radionuclides into the environment can be minimized. With time, radionuclides produced by nuclear fission accumulate in the nuclear reactor fuel and slow or stop the fission reactions. At this point, the fuel elements must be removed and processed to extract the fission products. The disposal of these nuclear wastes presents considerable problems. Liquid and slurry wastes from reprocessing are stored in steel or concrete tanks until the radioactivity falls to acceptable levels for discharge and dispersal, often into the sea, but, as some of the nuclides are very long-lived, there could be a longterm problem caused by their accumulation, possibly far from the point of discharge. Solid materials, with low levels of radioactivity, may be mixed with cement or other immobilizing agents and sealed in steel drums for burial, but some radioactivity may still remain when the drums eventually corrode. High level materials present severe problems caused by the high temperatures resulting from the radioactivity, by the longevity of some of the products (half lives of thousands of years) and by their toxicity. The most promising method of disposing of them seems to be vitrification, i.e., conversion to a glass-like product, followed by burial at a great depth under the earth's surface Very low level discharges of radionuclides may be made directly into sewers or into the atmosphere from high chimneys.

Radioactivity

Radioactivity is toxicant in the sense that it causes harm to living organisms. Most harm is caused by toxizing radiation.

History

This radiation modifies or inactivates biological molecules, frequently with fatal results.

The first observation of radioactivity was made in 1896 by Becquerel, who noted that salts of uranium were phosphorescent and emitted radiation which penetrated black paper, opaque to light, and which reduced a photographic plate. For pure salts this activity is independent of the mode of chemical combination and of changes in temperature and magnetic flux. The activity always relates to the amount of the source element present and hence radioactivity is described as an infra-atomic property.

A principal ore of uranium is pitchblende which includes its black oxide U_3O_8. Marie and Joliot Curie with Bemont noticed that the intensity of emission varied with the source.

Type of radioactive emission

The α–particles emitted by radium and uranium have the mass of helium alone but carry two positive charges ; one gram of radium emits 3.7×10^{10} α–particles and 2×10^9 cal, which is 2×10^5 times the calorific value of coal. The other principal source of radioactivity is the β-particle which is an electron carrying a single negative charge. Radioactive decay may give rise to other entities including the positron but these are not of concern as decay series (Fig. 5.1).

235 U is the predominant naturally occurring form, it loses 4 mass units and two nuclear charges on emission of an α-particle to give an isotope of thorium written as ^{234}Th. This product decays with the ejection of a β-particle, which entails no mass change but a gain of one nuclear charge to form protactinium (Pa) ; a further β-emission leads to ^{234}U. This isotope contributes only 0.7% of natural uranium and this can be accounted for by considering the half-lives ($t_{1/2}$) of its predecessors. The two β-emitters have very short lives relative to that of ^{238}U, while they sustain the sequence through their activity, their rate of formation is less than their rate of decay and so only small amounts survive in the equilibrium. Because $^{234}_{92}$U decays much more slowly than $^{234}_{91}$Pa, it will

accumulate to a relatively higher level than this precursor, nevertheless $^{234}_{92}U$ with $t_{1/2} = 2.5 \times 10^5$ yr is shorter-lived than $^{234}_{92}U$ by over 10^4 times and so survives at a lower concentration than the parent uranium.

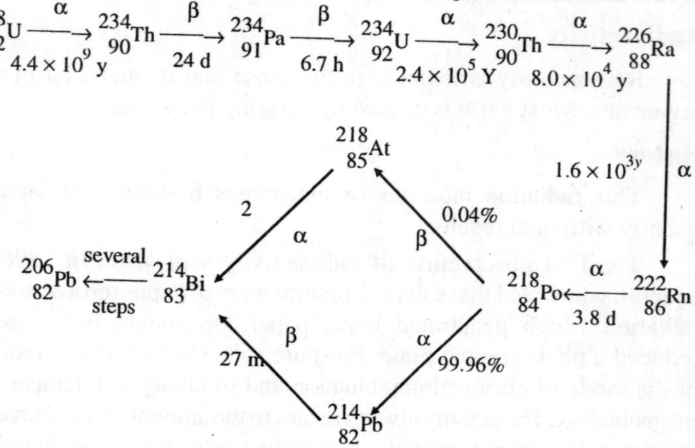

Fig. 5.1 Decay of uranium 238

The series is continued by the successive emission of two α-particles to produce ^{230}Th and then ^{226}Ra with eight fewer mass units than ^{234}Th and atomic number 88. Radium is in group II of the periodic table with chemistry similar to barium ; loss of an α-particle gives $^{222}_{86}Rn$ in the inert gas group zero. ^{218}Po in group VI is formed by further α-loss but this isotope has a dual decay path and yields the more electronegative ^{218}At by β-loss and also $^{142}_{82}Pb$ by α-loss. Radioactive decay ends after several further steps with the stable isotope ^{235}U and the other with ^{232}Th.

An important member of the ^{238}U series is the short-lived inert gas radon ($t = 3.8$ days) and owing to its release from granitic rocks we shall have to consider its toxicity within buildings erected on these foundations.

γ-Radiation. When a new species is produced by the emission of either an α-or β-particle it will be at an energy level above its stable ground state and to achieve stability the excess energy is emitted as γ-radiation.

Units of energy and measurement of toxicity

Energy is measured in electron-volts (eV). One eV is the energy acquired when unit electronic charge is accelerated through a potential difference of one volt and is equivalent to 1.60×10^{-12} ergs. The emission energies of radioactive particles are very high and are conveniently expressed in MeV or 10^6eV. β-particles have a range which depends on the emission energy ; for those of 0.5 MeV it is about 1 m and for those of 3.0 MeV it is about 10 m in air. When taken up by obstacles they produce a more penetrative secondary radiation known as bremstrahlung. The best protective barriers are solids of low atomic number such as aluminium, perspex and rubber. 2.5 cm of perspex will protect against β-emission of energy as high as 4 MeV.

Exposure to radiation is measured in rads where one rad leads to the release of 100 ergs/g of body tissue. This is also equivalent to exposure to a level of one Roentgen.

The current usage is to refer to exposure in rems. The rem takes account of a quality factor which for the more dangerous α-particles is 10, so that exposure to them at the level of one rad is rated as ten rems.

γ-rays are more difficult to stop because they are not ionic but their intensity falls away exponentially and a shield must be thick enough to reduce the intensity to acceptable levels.

Materials of high atomic number may exhibit photoelectric absorp-tion whereby the energy of γ-rays is transferred to an electron, which is then ejected from the atomic shield. ^{60}Co is a dangerous product of nuclear fission and emits γ-rays of mean energy 1.26 MeV ; to reduce their dose tenfold 4.6 cm of lead is needed and for each subsequent tenfold reduction a further 4.6 cm of shielding is necessary.

Nuclear fission

Discovery. Irene and Frederic Joliot-Curie came close in 1934 when they noted that the product from neutron bombardment of $^{238}_{92}$U was not the expected $_{89}$Ac but was more like $_{57}$Lc. Hahn and Strassman, while attempting synthesis of transuranium elements in 1938, were led to the first firm description of a fission reaction :

$$_{92}\mathrm{U} + {}^{1}_{0}\mathrm{n} \rightarrow {}_{56}\mathrm{Ba} + {}_{36}\mathrm{Kr} \qquad (1)$$

In 1940 McMillan and Abelson observed that some deuterons recoiled with high energy resulting from fission when they were used to bombard uranium, but a new element was formed (equation 2) which lacked the energy to escape from the body of uranium. This was named neptunium (Np) after the first planet beyond Uranus :

$$^{238}_{92}\mathrm{U} + {}^{2}_{1}\mathrm{H} \rightarrow {}^{238}_{93}\mathrm{Np} + 2{}^{1}_{0}\mathrm{n} \qquad (2)$$

Subsequently Seaborg showed that :

$$^{238}_{93}\mathrm{Np} \xrightarrow[2.1\text{ days}]{\beta} {}^{238}_{94}\mathrm{Pu} \qquad (3)$$

This isotope of plutonium, named after the furthest planet, is an α-emitter with $t_{1/2} = 86$ years ; that required for nuclear energy generation is $^{239}_{94}Pu$. *Criteria for fission.* Slow fission is possible but such events are rare and of no use for power generation, thus $^{238}_{92}U$ undergoes spontaneous fission to give $_{54}Xe$ which accumulates in natural sources and which can be used for dating.

$E_{critical}$ is that required to split a nucleus in two. When this energy is less than the binding energy the atom only requires to capture a neutron of zero energy to undergo fission ; it is said to be *fissile.* $^{235}_{92}U$ is the only natural material with this property, capturing a neutron and undergoing fission as $_{236}U$. The synthetic nuclide $^{239}_{94}Pu$ is also fissile.

A nucleus is said to be *fissionable* if it reacts only after capturing an eneretic neutron. In order to sustain an energy-generating chain reaction in fissionable nuclei one energetic neutron must survive to initiate the next generation. In practice more than one such neutron must be produced on fission since some escape from the system and others are captured by accumulating non-fissionable products.

The availability of energetic neutrons is expressed as the *multiplication factor* (k) = number of fissions in one generation/number in the preceding generation. Provided that $k > 1$ the conditions for maintaining a chain reaction will persist and be enhanced with each generation. This *supercritical state was necessary in the* $^{235}_{92}U$ bomb which was detonated over Hiroshima on 7 August 1945.

When $k = 1$ the system is said to be *critial* and this is the state required for the controlled release of energy.

When $k < 1$ the system is *sub-critical* and unproductive.

Quantitative yield. To appreciate the potential of fission for energy production it is helpful to analyse a typical reaction :

$$^{235}_{92}U + ^{1}_{0}n \rightarrow ^{95}_{42}Mo + ^{139}_{57}La + 2^{1}_{0}n + \beta\text{-particles} \quad (4)$$

$$\underbrace{235.0439 \quad 1.0087}_{236.0526} \qquad \underbrace{94.9057 \quad 138.9061 \quad 2.0174}_{235.8292}$$

By subtraction a mass loss Δm of 0.2234 units is obtained. To obtain the energy equivalent of mass the Einstein formula $E = mc^2$, where c is the velocity of light, 3×10^{10} cm/s. is used.

Using the formula 1 atomic mass unit = 931 MeV of energy It follows that 1 g of $^{235}_{92}U$ will produce 2.38×10^{24} MeV of energy .

Preparation of the fuel

The problem of enriching the low level (0.7%) of $^{235}_{92}U$ in natural uranium was initially dependent on conversion of the oxides to uranium hexafluoride. Then the corrosive mixed gases were passed through nickel columns with pores of 0.01 μm when the lighter $^{235}UF_6$ escaped faster than $^{238}UF_6$; since the fractional separation was only 1.002 a multi-stage process was needed. After enrichment the hexafluoride was converted to the dioxide by hydrolysis :

$$UF_6 + \text{steam} \rightarrow UO_2.F_2 \xrightarrow{H_2} UO_2 + 2HF \qquad (5)$$

Alternatively the tetrafluoride may be reduced to uranium metal

$$UO_2 + 4HF \xrightarrow[-H_2O]{550^\circ C} UF_4 \xrightarrow[700^\circ C]{Mg} U + 2MgF_2 \qquad (6)$$

A more economical process is to place the UF_6 in a series of gas centrifuges where now the heavier isotope migrates more readily to the boundary and $^{235}UF_6$ becomes concentrated near the axis of rotation. *Laser separation* depends on the raising of uranium metal to an excited state rather in the manner of AA spectroscopy A laser emitting green light can be tuned precisely to a wavelength of 502.73 nm so as to excite only ^{235}U. It is then ionized by a second source and electromagnetically separated from residual uncharged ^{238}U. In practice much of the original uranium vapour is also collected and so the net change is one of enrichment and not complete separation.

Nuclear reactor types

The pressurized water reactor (*PWR*). This is widely used today around the world. Ordinary water is used as a moderator for the production of low energy or *thermal* neutrons, that is those with energy comparable to the system as a whole. Water is favoured because it is cheap and being composed of lighter atoms it is efficient. The oxide fuel containing 3% $^{235}UO_2$ is supplied as 1 cm pellets packed within 3.5 m long hollow zircalloy tubes or pins, which are grouped into bundles or elements of about 20 c m sup 2 . The reactor contains about 200 of these elements with spaces for control rods, coolant flow and neutron detecting instruments. Moderating water enters at 290°C and leaves at 320°C and hence the reactor chamber is at a high pressure of about 2000 psi or 15 MPa. The containing steel vessel is some 12 m high and 5 m in diameter.

The pressurizer is fitted to ensure that there is no loss of coolant by evaporation ; this would be especially serious as neutron energies would also rise wherever steam replaced water.

The issuing hot water is passed through a heat exchanger to produce steam which leaves to drive the turbine. After exchanging its heat the cooling water is returned at, while fesh cooling water and turbine condensate enter the exchanger at. The primary coolant in contact with the reactor is isolated from the secondary heat exchanger circuit.

The multiplication factor may be rapidly varied by movement of the control rods which are withdrawn to raise the power level and inserted to reduce it. If need be they can shut the reactor down completely. The rods are neutron absorbing materials such as boron steel or silver alloys and their action is supplemented by injection of boric acid solution into the moderator.

The boiling water reactor (*BWR*). At pressures of about 7 MPa the reactor can be stabilized even when the moderator boils and the emitted steam can be used directly to drive the turbines. However, these and all the linking condensers and pipe work must also be shielded as the circulating water becomes radioactive.

The heavy water reactor. The Canadian design is the CANDU, an acronym for Canada and Deuterium. It was developed there after World War II to take advantage of available natural uranium without the need for enrichment ; this is because heavy water has a low absorption for thermal neutrons. It also follows that to reach thermal energies energetic neutrons must undergo more collisions and so travel further in D_2O than in light water.

The fuel rods pass through a tank, or calandria, containing the moderator at a low pressure while a high coolant pressure is maintained only within the fuel channels. This has the advantage of limiting the volume of the vessel but it also means that the steam pressure and temperature are less than in the PWR and the efficiency is only 30%.

Breeder reactors the liquid metal fast breeder reactor. These reactors are attractive to nations with no indigenous fuel supply because they produce fissile material in the course of their operation and can use depleted fuel from thermal reactors.

Here the driving force is obtained from fast neutrons and there is therefore no moderator ; the cooling and heat transfer circuits are charged with metallic sodium (m.p. 98°C). From a purely chemical point of view this is a hazardous choice because of its high reactivity to water and oxygen and the container must be flooded with nitrogen gas to limit these risks. However, it has engineering advantages in that sodium has only a slight moderating effect and at the operating temperature of 500°C a high pressure vessel is not required. In the breeder reactor sodium within the inner cooling circuit captures some neutrons to form the higher nuclide :

$$^{23}_{11}Na + ^{1}_{0}n \rightarrow ^{24}_{11}Na \qquad (7)$$

Hence a double looped system is required to prevent circulation of active material outside the screen.

If a breeder reaction is to be sustained, the ratio of fissile atoms produced to the average number of fuel atoms consumed must be > 2. This is because of the demands : one nuetron is required to sustain the critical state, one neutron is required to produce fissile material and some excess is necessary to make up losss.

Under typical conditions the reactor will be charged with ^{238}U mixed with about 20% by weight of ^{239}Pu. When ^{239}Pu captures a neutron of essentially zero energy the product ^{240}Pu has binding energy of 6.4 MeV which is in excess of the energy required for fission (4.0 MeV) and so the self-sustaining reaction begins. In a year of operation about 50 times the intial weight of ^{239}Pu will be produced together with other products (Table 5.1).

Table 5.1. Some fission products

Element	*Symbol*	$t_{1/2}$
Strontium	^{89}Sr	50 days
	^{90}Sr	28 years
Ruthenium	^{106}Ru	1 year
Caesium	^{134}Cs	2.1 years
	^{137}Cs	30 years
Iodine	^{129}I	1.7×10^7 years
	^{131}I	8 days
Plutonium	^{238}Pu	86 years
	^{239}Pu	2.4×10^4 years
	^{240}Pu	6.5×10^3 years
	^{241}Pu	13 years
	^{242}Pu	3.8×10^5 years

Strontium 90 presents a risk inherent in its relatively long half-life and its chemical affinity with calcium, which leads to its concentration

and persistence in bone ; particulate emission also presents a hazard to lungs.

Caesium 137 is similar to potassium in its chemistry and therefore it penetrates neural tissue and presents a hazard to muscle.

In the aftermath of the Chernobyl incident concern arose from the spread of ^{137}Cs.

Iodine 131 is combined and concentrated in the thyroid gland following the ingestion of milk from cows grazing on polluted pasture.

Ruthenium 106 has been found in seaweed at levels of 200 pCi/g – a concentration factor of 2000 with consequences for those who consume laverbread.

Plutonium nuclides do not follow a natural path in a food chain as they are artificial, nevertheless ingestion of their water-soluble forms presents a hazard to bone and to liver. Insoluble forms inhaled as dust induce lung cancer.

Measuring Radiation

The systems for measuring radiation and radiation damage are complex. This complexity is not simplified by the fact that there are two measurement systems in existence – the traditional method, and the newer SI (System Internationale) method. If the two systems are used together, it is rather like talking in feet and inches and metres and centimetres. This discussion is confined to the SI system, and where the old system is quoted, equivalents are given.

When a radioactive substance decays, it undergoes a number of decays each second. In the SI system, one *becquerel* refers to one disintegration per second of any radionuclide ; 3,000 becquerels means that there are 3,000 decays per second.

This is merely a system for counting the number of atomic events, however, and does not indicate how harmful this will be to human beings. To assess this, a system has been developed which looks at the effects of radiation. The amount of radiation energy absorbed in living matter is called the *dose.* This can be received from one radionuclide or several, inside or outside the body. The dose is measured in units called *grays.* One gray equals one joule of energy absorbed by one kilogram of tissue.

A dose of alpha radiation is about 20 times more damaging than the same amount of beta or gamma radiation. The dose needs to be adjusted to take this into account, and the final calculation is referred to as the *dose equivalent,* which is measured in *sieverts.*

Other terms which may be useful to understand are :

- the *effective dose equivalent,* also measured in sieverts, which calculates the effect of a given radiation dose to different parts of the body (the breast, say, being more vulnerable than the thyroid, the dose is multiplied by a weighting factor to take this vulnerability into account) ;
- the *collective dose equivalent* describes the dose equivalents received by a number of people, rather than just one person.

Equivalents, old and new systems

1 Curie = 37,000,000,000 ; 100 rad = 1 gray beequerels 100 rem = 1 sievert

1 rem = 10 millisieverts

1 millirem = .01 millisieverts (10 mierosieverts)

RADIATION EFFECTS

When ionizing radiation from a radionuclide hits an atom it strips off an electron and releases a large amount of energy. Thus, almost any biological molecule hit directly by ionizing radiation is functionally destroyed. This, in itself, is enough to cause considerable damage to a living cell but the damage is accentuated by the formation of highly reactive free radicals. Since water is the main component of any cell, most of these radicals are either H or OH. Radiation damage of any kind is greatest in the presence of oxygen.

Direct damage to DNA, or indirect damage from reaction with free radicals, causes chromosome breaks, cross-linking of the polynucleotide chains or modification of the constitutent bases. Various repair processes can occur but there is still a high probability of mutations being produced. Most of these mutations are lethal and the rest are frequently teratogenic or carcinogenic.

Damage to proteins is associated with a general drop in enzyme activity which has been attributed mainly to the splitting of disulphide bonds. Hydrolytic enzymes may appear to :

- There is no known tolerance level for ionizing radiation, i.e. any dose is harmful. The extent of harm depends on the sensitivity of the individual and the amount of the dose received.
- Most radiation received by the population at large comes from natural background sources.
- Most artificial radiation received by the population at large comes from X-rays.

- An additional, though by comparison small, radiation loading is imposed on the public by the use of nuclear weapons and nuclear power.
- Exposure to radiation should be treated as a risk, to be quantified against the benefits to be gained by such exposure. Any unnecessary exposure to radiation should be avoided.

A **tolerance (or threshold) level** is a level below which no damage occurs.

The effects of a high dose of radiation to the whole body are a matter of scientific agreement : death. The United Nations Scientific Committee on the Effects of Atomic Radiation has determined the effects of different whole-body doses, which are listed below :

- Doses of 100 grays (*100 sieverts*) :
  death through central nervous system damage in hours or days.
- 10 to 50 grays (*10 to 50 sieverts*) :
  death through gastrointestinal damage in about one to two weeks.
- 3 to 5 grays (*3 to 5 sieverts*) :
  death for half of the people exposed in one to two months (bone marrow damage).

Effects of lower doses are :

- 150 to 250 rems (*1.5 to 2.5 sieverts*) :
  nausea, vomiting, probable skin burns, foetal or embryonic death if pregnant ; long term, people in ill-health may not survive, people in good health will apparently recover, though with possible permanent health damage.
- 50 to 150 rems (*0.5 to 1.5 sieverts*) :
  less severe radiation sickness and burns, spontaneous abortion or stillbirth if pregnant ; long term, possible benign or malignant tumours, premature ageing, shortened lifespan, genetic damage to offspring.
- 10 to 50 rems (*0.1 to 0.5 sieverts*) :
  most people experience little immediate reaction ; sensitive people may have radiation sickness ; long term, possible premature ageing, genetic effects and some risk of tumours.
- under 10 terms (*under 0.1 sieverts*) :
  no immediate effects ; long term, premature ageing, some offspring mutation, risk of tumours.

Different Tolerances of Radiation

10 sievert	(= 10,000 millisieverts)	death
below 1 sievert	(= 1,000 millisieverts)	no 'early' death
below 0.1 sievert	(= 100 millisieverts)	no radiation sickness in sensitive individuals
0.03 sieverts	(= 30 millisieverts)	one X-ray
0.001 sievert	(= 1 millisievert)	one year's dose from backgrond radiation
0.0001 sievert	(= 0.01 millisievert)	one modern X-ray

All the figures, excepting the figure for background radiation refer to whole-body doses over a short period of time. If the dose is spread over a longer period, the body is better able to cope. Because of the natural background radiation that we are all exposed to in our everyday lives, our bodies have developed repair mechanisms for radiation damage.

According to the UK Atomic Energy Authority, a single dose of 4 sieverts would result in a one in two chance of death : the same dose delivered gradually over a year would probably be tolerated because of the body's natural repair processes, although with possible long-term consequences.

There is a variation in radiation sensitivity between the different organs of the body. Adult tissues are relatively robust : the kidneys, bladder and cartilage can take relatively high levels of radiation, though red bone marrow, eyes, testes and ovaries are more sensitive.

Sensitivity differs between individuals. Some adults, and all children, are particularly at risk. Quite small doses of radiation to a child's cartilage can slow or halt the growth of bones and lead to deformity. The younger the child, the more severe the stunting. Irradiation of children's brains during radiotherapy can cause changes of character, loss of memory and, in very young children, dementia and idiocy.

Unborn children are particularly prone to brain damage, if their mothers are irradiated between the eighth and fifteenth week of pregnancy. This danger extends to X-rays, which can cause severe mental retardation to an unborn child.

Large doses of radiation are given only to patients who are already so ill that the risk of the treatment is justified.

Dangers of Radiation

Scientific consensus now exists concerning the effects of large doses of radiation. X-rays and radioactivity were discovered in the 1890s' radium was regarded as having very strong medicinal qualities, and was sold for a number of miracle cures. Its harmful qualities were first noticed by Henri Becquerel when he put a vial of radium in his pocket and damaged his skin. Nevertheless, the dangerous qualities of radiation were not fully understood, and at least 336 early radiation workers died as a result of the doses they received – including Marie Curie, the discoverer of radium.

As more has been found out about radiation, so safety standards have become strictor. Permissible *occupational exposure* to ionizing radiation in the USA has been tightened up over the years (equivalents are now in millisieverts, 1,000 millisieverts = 1 sievert) :

- 52 roentgen (470 to 520 millisieverts) per year in 1925 ;
- 36 roentgen (320 to 360 millisieverts) per year in 1934 ;
- 15 rem (150 millisieverts) per year in 1949 ;
- 5 – 12 rem (50 – 120 millisieverts) per year in 1959.

The allowable dose for *public* exposure is much lower than 50 millisieverts.

Potential Radiation Risks

Nuclear power stations, the reprocessing of nuclear fuel and the disposal of nuclear waste, all result in the emissions of various levels of radioactivity.

- The radioactivity emitted by an efficiently-run nuclear power station is small compared to natural background radiation although people in the immediate vicinity will receive a larger radiation dose from the plant than people living further away.
- The total radioactive load to the planet is much larger when the effects of digging up radioactive raw material, and disposing of nuclear waste, are taken into account.
- There is a risk of a large release of radioactivity from a nuclear accident, act of sabotage or bombing during a conventional war.
- Releases of radiation from nuclear accidents are often belittled by expressing them as a percentage of natural background radiation. There is no scientific basis for this, Every dose of radiation is harmful, so additional doses cannot be justified by comparing them to existing levels.
- Atmospheric radiation from testing nuclear weapons is a small percentage of overall radiation levels. It has had serve localized effects, however, with global implications which will persist for hundreds of thousands of years.

Risks and Benefits of Radiation Exposure

Exposure to radiation should be treated as a risk, to be quantified against the benefits to be gained by such exposure. Any unnecessary exposure to radiation should be avoided.

This statement is accepted by the international bodies who set radiation protection limits.

- Radiotherapy, in which large doses of radiation are given to treat serious illnesses such as cancer, is accepted as a reasonable exposure to radiation. The risk is large, but in many cases the risk of not receiving the treatment is larger.
- The use of X-rays for medical diagnosis is also accepted as valid, provided that they are used for specific reasons and not as, say, a screening system for new employees in a firm.
- Natural background radiation is a variable about which little can be done, although a debate is opening as to how 'natural' should be defined.

SAFETY STANDARDS

Safety standards for exposure to radiation are set by the International Commission on Radiological Protection (ICRP). In 1952, the ICRP issued its recommended exposure limits to radiation, which was the same standard agreed by nuclear physicists after the Second World War 5 rem (50 millisieverts) to the whole body.

The revised limits in 1959 recommended that workers'5 rem (50 millisieverts) whole-body exposure should be calculated by adding external and internal exposure together (internal exposure occurs by swallowing or breathing in radioactive materials). The limits have stayed the same, except for an upward revision in 1977 of exposure to certain parts of the body, to the present day.

Exposure limits for radiation radiation workers are based on the assumption that they know the risks of what they are doing, and are healthy, fit, and capable of withstanding a relatively high exposure. The ICRP recommended that the public, for whom greater protection is necessary, should receive no more than one-tenth of the occupational exposure, i.e. a whole-body dose of 0.5 rem (5 millisieverts) over a year.

The ICRP accept that there is no threshold limit for damage from radiation exposure, and that all doses can cause some damage to the body. The 5 millisieverts exposure limit for the general public was considered the absolute maximum, while a limit of 1 millisievert was considered a better value to adopt. West Germany decided on a 0.35 millisievert limit, the USA took 0.25 to be its limit – but the UK adopted a limit of 5.0 millisieverts.

REFERENCES

Radionuclides

Allaby, M. (ed). (1988) *Dictionary of the Environment,* MacMillan, Basingstoke. Burn, D. (1978) *Nuclear Power and the Energy Crisis,* MacMillan, Basingstoke. Eisenbud, M. (1987) *Environmental Radioactivity from Natural, Industrial and Military Sources* (3rd edition), Academic Press, Orlando.

Lilienthal, D. (1980) *Atomic Energy : a New Start,* Harper and Row.

Marshall, W. (ed.) (1983) *Nuclear Power Technology,* Vol. 3, *Nuclear Radiation,* Clarendon Press, Oxford.

Nuclear Installations Inspectorate (1992) *Safety Assessment Principles for Nuclear Plants,* NII, London.

UK Royal Commission (1976) *6th Report. Nuclear Power and the Environment,* HMSO, London.

Yearbook of Science and Technology (1992) McGraw-Hill, p. 297

CHAPTER 6

Pesticides

Pesticides are a variety of chemical substances used to control organisms which may adversely affect public health, or organisms which attack food and other material essential to mankind. Such organisms include rodents, insects, nematodes and fungi. Chemicals used against infectious bacteria causing human animal or plant diseases as well as those employed against viruses, protozoa and internal parasites of animals plants and humans are generally not classified as pesticides.

Pesticide Types

There are many types of pesticides: **insecticides** which kill all or specific species of insects; **herbicides** used against weeds; **fungicides** against fungi; **acaricides** against mites; **nematicides** against nematodes; **rodenticides** against rodents; **molluscides** against smails and slugs.

The Growth of the Use of Pesticides and Toxicity

Modern pesticide research started in the last century, with the development of an insecticide which checked the spread of Colorado beetle in the USA. The rapid growth of pesticide research occurred before and during the Second World War, and the current generation of pesticides have been developed within the last 40 years.

Pesticides have made possible great increases in food production and improvements in human health. Where harm has occurred, this has largely been due to ignorance of the properties of these chemicals and misuse, i.e., wrong quantity, application or timing. Since it is impossible to know every property of every pesticide it is inevitable that some environmental damage will occasionally result from their use. However, this is no argument against their use, so long as the benefits can be clearly shown to outweigh the damage. The important thing is to be aware that there is a finite risk of damage, to minimize it by appropriate toxicity testing of chemicals before they are applied on a large scale and to ensure that the application is carried out correctly. When pesticides are applied on a large scale, a constant watch must be kept for environmental damage so that corrective steps may be taken as soon as possible.

Misuse of pesticides can often lead to major problems. Wearing protective clothing—in some cases respirators in others aprons, boots and face masks—are a must as pesticides enter the skin through feet and hands if left unprotected. This could lead to beefy dermatitis. Containers used to store pesticides should never be used for storing grain.

All pesticides should be treated as what they generally are—dangerous substances. This includes pesticides sold for garden or home use. These international hazard symbols will often give useful information about a particular product :

HARMFUL

TOXIC

OXIDIZING

FLAMMABLE

EXPLOSIVE

CORROSIVE

Pesticides and herbicides have made possible great increases in food production and improvements in human health. Where harm has occurred, this has largely been due to ignorance of the properties of these chemicals and misuse, *i.e.*, wrong quantity, application or timing. Since it is impossible to know every property of every pesticide or herbicide, it is inevitable that some environmental damage will occasionally result from their use. However, this is no argument against their use, so long as the benefits can be clearly shown to outweigh the damage. The important thing is to be aware that there is a finite risk of damage to minimize it by appropriate toxicity testing of chemicals before they are applied on a large scale and to ensure that the application is carried out correctly. When pesticides are applied on a large scale, a constant watch must be kept for environmental damage so that corrective steps may be taken as soon as possible.

Everyday cases of poisoning by pesticides in developing countries are reported. In 1975, hundreds of people in Karnataka's Shimoga district in India were struck by a mysterious attack of arthritis which wastes limbs and brings about dwarfism. Studies indicate that pesticides that were sprayed in fields were ingested by crabs which, in turn, were eaten by farmers, poisoning them as well. According to an WHO estimate about 750,000 people are poisoned by pesticides every year, 14000 of which are fatal. Three-fourths of these occur in the third world. The use of pesticides has been increasing at snow-balling rates. In India, pesticide consumption rose from 2000 tonnes in 1955 to 80,000 tonnes in 1984—a 40-fold increase. By the end of the current year, the estimated consumption is projected at 100,000 million tonnes.

The scale of ecological havoc being wrought by pesticides is already alarming. When a pesticide is sprayed, much of it reaches the soil. Water becomes contaminated from direct application or run-off. Minute quantities become airborne. Many chemicals become metabolized by animals that eat plants sprayed with pesticides. We, thus, find these toxic chemicals into the human food chain.

Misuse of pesticides can often lead to major problems. Wearing gloves and face masks are a must as pesticides enter the skin through feet and hands if left unprotected. This could lead to beefy dermatitis. Containers used to store pesticides should never be used for storing grain.

INSECTICIDES

Organochlorines

Organochlorine compounds also known as Chlorinated hydrocarbons are the most prevalent pesticides in the environment because of their wide use and persistence. The most persistent of these compounds are DDT and derivatives such as DDD and DDE. DDE has an environmental half life of ten years or more, DDE can persist for decades. Almost as persistent are lindane (BHC) and heptachlor. Less persistent are aldrin and dieldrin, but even these require 2 years in the soil for 95 per cent degradation. All of the compounds mentioned are currently in use throughout the world, though DDT, aldrin and dieldrin have been banned for most purposes in the United States, and their use has been discouraged in the United Kingdom and a number of other countries. Reaction against the use of these compounds has followed awareness of their toxicity to nontarget organisms and the discovery that some target insect populations were becoming resistant. However, substantial quantities of chlorinated hydrocarbons are still used in wood preservation and where there is, at present, no satisfactory substitute. The most extensive use of DDT now occurs in tropical countries because it is cheap, persistent, generally effective and with minimum harm to human beings. Until a suitable alternative is available, it use will continue. At present, possible substitutes are much more expensive, less presistent and more toxic to human beings.

DDT (dichlorodiphenyl trichloroethane)

Apart from the ready availability of chlorine and petroleum products in the years following the Second World War, the successful use of DDT in controlling disease during its final stage encouraged the research and development of organochlorines as a whole. DDT was first described by Othmar Zeidler in 1874, although over sixty years elapsed before it was formulated as an insecticide by the Geigy company.

The preparation of DDT depends on the condensation of chloral with chlorobenzene in sulphuric acid (Fig. 6.1)

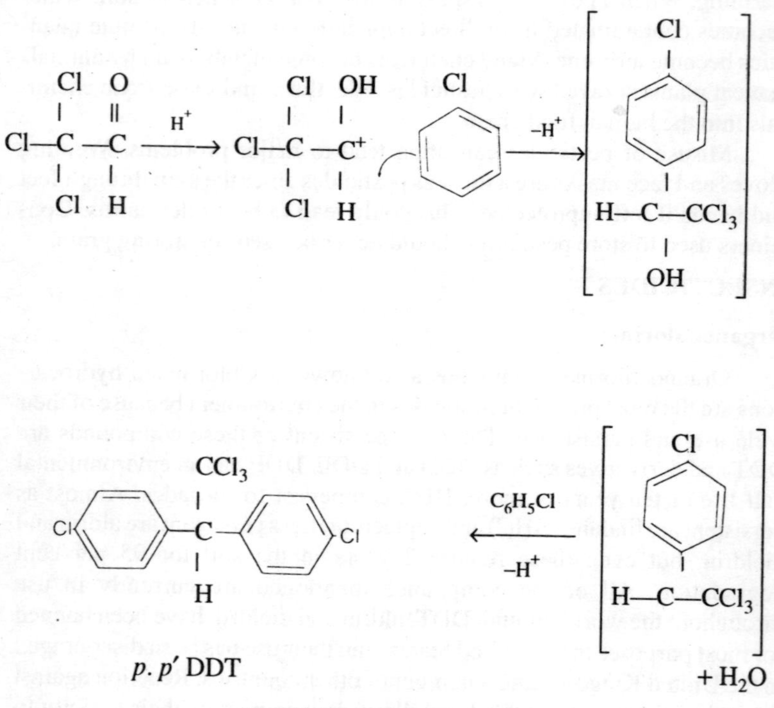

p. p′ DDT

Fig. 6.1. The preparation of DDT

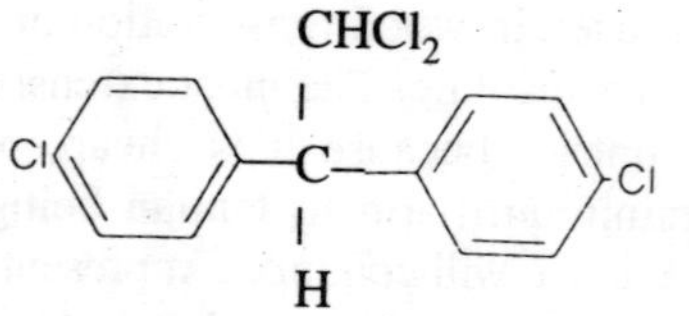

Fig. 6.2. Structure of DDD

Significant amounts of the isomer with one chlorine in an *ortho*-position are produced; also chloral produced by the chlorination of acetaldehyde is contaminated by dichloroacetaldehyde and gives rise to DDD. Commercial DDT has been shown to be a mixture of 14 substances.

Lindane (hexachlorocyclohexane)

This substance is obtained by the addition of three molecules of chlorine to benzene activated by ultraviolet irradiation. In theory there are eight possible geometrical isomers in which the chlorine atoms occupy different relative positions about the cyclohexane ring. Structure shown in Fig. 6.3 is drawn in a way which shows the characteristic 'chair' form with six axial bonds parallel to the main axis and six equatorial bonds disposed around the belt of the molecule. The active form, the *gamma*-isomer, has three consecutive axial and equatorial substituents and forms about 15% of the mixture of reaction products which includes five of the possible isomers. It was first introduced commercially by ICI in 1942.

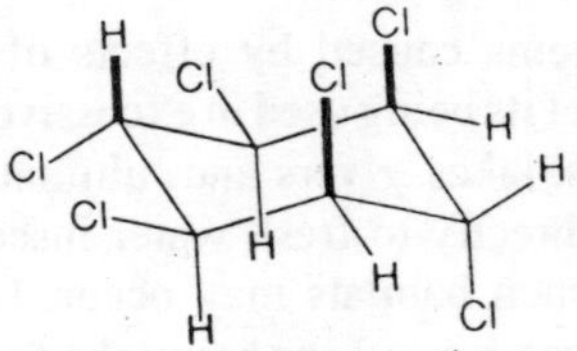

Lindane

Fig. 6.3 The Structure of Lindane.

Lindane or gammexane has similar properties to DDT and is widely used in Third World countries as the crude product is cheap to produce. For control of pests in food crops such as the potato it is necessary to purify lindane, which is tasteless, by recrystallization. It is superior to DDT in controlling soil pests.

Some other chlorinated pesticides

The structures of some other principal chlorinated pesticides are shown in Fig. 6.4. A number of these compounds were obtained by the Diels-Alder addition reaction of hexachloropentadiene. Combination with bicycloheptadiene yields Aldrin, which on epoxidation forms Dieldrin.

Epoxidation is an important in *vivo* process and results in the hydroxylation and consequent solubilization of organic pollutants. While this change is the key to degradation of toxic compounds there are occasions when the cure is worse than the disease. Thus Aldrin is converted into Dieldrin *in vivo* and the latter is in general more toxic to fish. The concentration of residual chemicals in food chains is a further cause for concern and has led to restrictions on the use of these compounds.

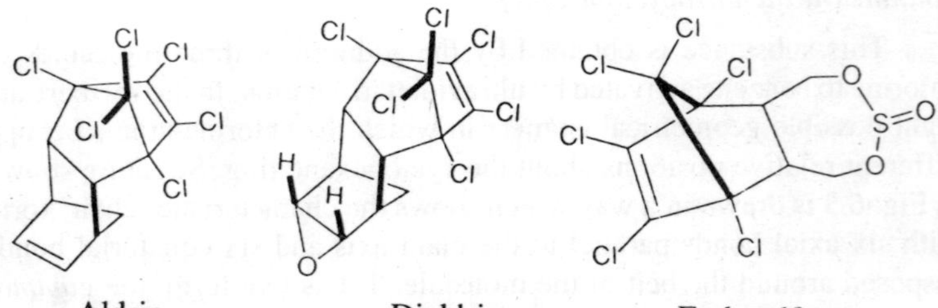

Fig. 6.4 The Aldrin group of pesticides.

The cyclic sulphite Endosulfan is still in use, since the high level of oxygenation ensures that it is less persistent than other members of this group.

Toxic Effects of Organochlorine Pesticides

Most of the problems caused by effects of DDT on non target organisms are the result of its being used in excessive amounts. The excess finds its way into ponds, lakes, rivers and, ultimately, the sea. In some cases, DDT is applied directly to fresh water insect habitats; in others, accidental spraying to such habitats may occur. Otherwise, DDT may enter waterways in surface run-off, or be washed out of the atmosphere inn rain or snow. Despite this, the concentration in natural waters is low. However, many plants and animals tend to accumulate DDT and its more persistent derivative DDE (Fig. 6.5). For example, oysters can concentrate DDT from 1 ppb in seawater to 700 ppm in their bodies. Water fleas can concentrate DDT from 0.5 ppb in the surrounding water to 50 ppm in their bodies. This process of accumulation can continue through food chains until harmful levels are reached in the ultimate predators. The effect is, of course, greater in the case of DDE owing to its greater persistence and prevalence. Accumulation occurs largely because of the affinity of DDT and DDE for fats. This leads to their localization in animals in adipose tissue where the turnover rate is low. Stress may cause mobilization of adipose tissue. The subsequent release of the accumulated residues may lead to toxic levels in the blood of animals even after exposure to the pesticides has ceased.

The harmful effects of chlorinated hydrocarbons were first noted in birds of prey where reproductive failure led to a marked decline in many populations. This was partly due to impaired calcium metabolism which resulted in fragile eggs with abnormally thin shells, and partly due to behavioural changes which favoured egg breakage. This has proved to be a problem in battery chicken farming too, where sawdust from wood treated with chlorinated hydrocarbons has been used as litter. Harmful effects on fish have also been noted, including behavioural changes leading to reproductive difficulties increased mortality among the young and, in some cases, acute toxicity to adults. Further, small concentrations of DDT (0.01 ppm) have been shown to reduce photosynthesis in marine plankton, while even smaller concentrations (1 ppb) in seawater can kill many brine shrimps (*Artemia salina*) within a few weeks.

Agarwal (1983) has given an account of pesticide pollution of waters and has compiled data on mean levels of DDT residue in some worked-out river and lake waters of India. During 1976-78, the DDT residue in River Yamuna at Delhi was 0.249 ppm in upstream and 0.558 ppm in the downstream off Wazirabad.

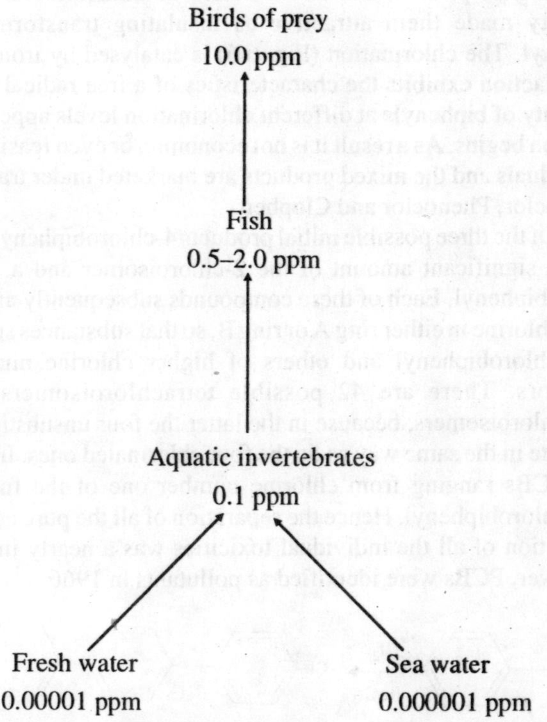

Fig. 6.5 A simplified diagram showing food chain accumulation of DDT.

Another insecticide in this group which has caused environmental damage in the USA is kepone. This compound was used domestically as an ant or cockroach poison, but its production was stopped when it was shown to have caused brain and liver damage, sterility, slurred speech, loss of memory and eye-twitching in workers who had been exposed to it. Matters were made worse by leakage of kepone from the production plant into the main sewer system of Hopewell, Virginia USA and, therefore, into the James River. As a result, fish and shellfish stocks were contaminated and the Governor of Virginia closed more than 100 miles of the river to commercial fishing until such time as residue concentrations should fall to a safe level. Another insecticide, mirex, which is chemically identical to kepone except for one oxygen atom, is used to control ants, but there is concern about its possible carcinogenicity.

Polychlorobiphenyls (PCBs)

These compounds were first produced industrially in 1929 and were used in printing inks and paints. Later their use was extended to the softening of plastics and as insulators in transformers. Their thermal stability made them attractive as insulating transformer fluids and biphenyl. The chlorination (Fig. 6.6) is catalysed by iron (III) chloride. The reaction exhibits the characteristics of a free radical reaction and a diversity of biphenyls at different chlorination levels appear shortly after reaction begins. As a result it is not economic, or even feasible, to separate individuals and the mixed products are marketed under trade names such as Aroclor, Phenoclor and Clophen.

Of the three possible initial products 4-chlorobiphenyl predominates with a significant amount of the 2-chloroisomer and a little of the 3-chlorobiphenyl. Each of these compounds subsequently and rapidly captures chlorine in either ring A or ring B, so that substances such as 2,2',3,4' -tetrachlorobiphenyl and others of higher chlorine number occur in Aroclors. There are 42 possible tetrachloroisomers and also 42 hexachloroisomers, because in the latter the four unsubstituted positions permute in the same way as do the four chlorinated ones. In total there are 209 PCBs ranging from chlorine number one of the fully substituted decachlorobiphenyl. Hence the separation of all the pure components and evaluation of all the individual toxicities was a nearly impossible task. However, PCBs were identified as pollutants in 1966.

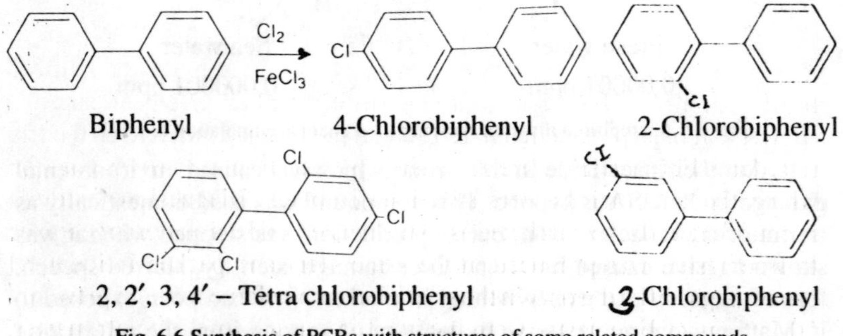

Fig. 6.6 Synthesis of PCBs with structures of some related compounds.

Toxic Effects of PCBs

Polychlorinated biphenyls (PCBs) are structurally similar to DDT and have similar effects. They are stable at temperatures up to 800°C resistant to acids, bases and oxidants, and only sparingly soluble in water. Because of these properties, many uses have been found for them. They may act as heat transfer fluids, insulators, hydraulic or plasticizers, and are incorporated in paints, adhesives, gasket sealers, brake linings, fluorescent lamps and carbonless copying paper.

Polychlorinated biphenyls enter the environment accidentally or through poor waste disposal. Like DDT and DDE, they are persistent in the environment because of their chemical stability. Also, their lipid solubility facilitiates food chain amplification. Their acute effects are not a serious problem but their chronic effects are so similar to those of DDT and DDE that its is likely that they act synergistically. Because of this, their use is now carefully controlled in most countries and largely restricted to closed system applications. At normal levels of exposure PCBs are not very toxic to man, although in the Yusho incident their inclusion in rice bran cooking oil led to serious consequences. Residents in South Japan became affected by a peculiar skin disease which first broke out in March 1968; it became notorious as 'Yusho disease'. Another outbreak occurred in Taiwan in March 1979 when about 2000 people were affected. In 1973 225—450 kg of the related polybrominated biphenyls (PBBs) were accidentally added instead of magnesium oxide to livestock feed in Michigan, USA. This resulted in contamination of farm animals making necessary the destruction of approximately 29,800 cattle, 5920 pigs, 1470 sheep and 1.5 million chickens. Also rendered unsaleable were more than 880 tonnes of feed, 8.1 tonnes of cheese, 1.2 tonnes of butter, 15.4 tonnes of dry milk products and nearly 5 million eggs. It is impossible to know how many people have suffered sublethal poisoning as a result of eating contaminated meat, eggs and milk, but it may be a very considerable proportion of the population of Michigan. Amongst the possible effects of poisoning with PBBs have been identified teratogenicity and carcinogenicity in rodents, and porphyria and reduced fertility in birds. The origins of this disaster seem to have been due to a shortage of bags prelabelled in red so that the PBBs, marketed as the fire retardant 'Firemaster', were for a time packed in plain brown bags on which the trade names were stencilled in black. Magnesium oxide, a normal feed additive, marketed as 'Nutrimaster' and similar in appearance to Firemaster, was produced in the same plant and packed in identical brown bags, differing only in the stencilled name. Somehow, ten to twenty 50 Ib bags of Firemaster were included in a truck-load of Nutrimaster.

Because of the superficial similarity of contents and packaging this was not noticed. As a result, animal and poultry feed throughout Michigan was contaminated. To make matters worse, some farmers sold contaminated animals to rendering plants to be processed into livestock feed, dog and cat food as well as tallow and grease used in soaps, shampoos etc. This accident illustrates the importance of clear labelling of toxic substances and of constant vigilance to spot poisoning before it becomes widespread. Once the distribution of PCBs and the similarity of toxic effects to those of DDT were established, their use declined and was banned. Production worldwide had practically ceased by 1977 but the problem of destruction of the residues remained.

Organophosphorus Insecticides

The organophophate insecticides include malathion, diazinon and parathion. They are readily metabolized by mammalian enzymes and rapidly degraded in the environment. Thus, any environmental damage caused by these compounds will tend to be localized in the area of application. However, these compounds are related to the nerve gases developed for use in war and can be lethal to many different organisms, including humans. Hence, anyone applying these compounds should wear protective clothing, including goggles, rubber gloves and even a respirator. One should avoid treating area which are important habitals for wildlife.

The principal action of the organophosphate compounds is 01 inhibit acetylcholine esterase. Thus, the action of acetyl choline released at the nerve synapses ceases to be finite with each nerve impulse. Amongst the consequences are tremors in involuntary muscles, convulsions and death.

Parathion was one of the earliest in use in agriculture being effective against and red spider mite. The methyl ester was found to be preferable as it retained the activity against pests but with less risk to mammals, however this is still considerable and Parathion is increasingly being replaced by less toxic compounds.

$(C_2H_5O)_2P(=S)-O-C_6H_4-NO_2$

PARATHION

O, O - diethyl - O - 4 - nitrophenyl-phosphorothioate

$(CH_3O)_2P(=S)-S-CH(CO_2Et)-CH_2-CO_2Et$

MALATHION

O, O - dimethyl -S- 1, 2-(ethoxycarbonyl)-ethyl phosphorohioate

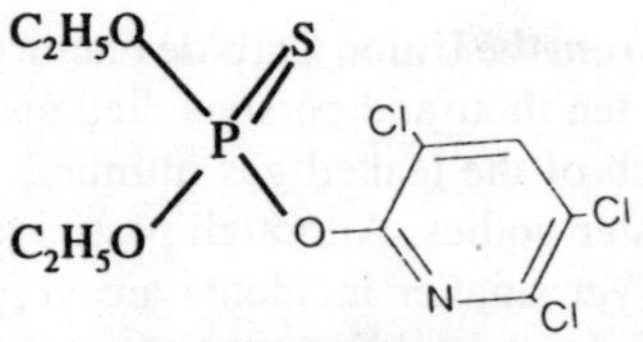

CLORPHYRIFOS
O, O -diethyl - O - 3.5.6 - trichloro - 2-pyndyl phosphorothioate

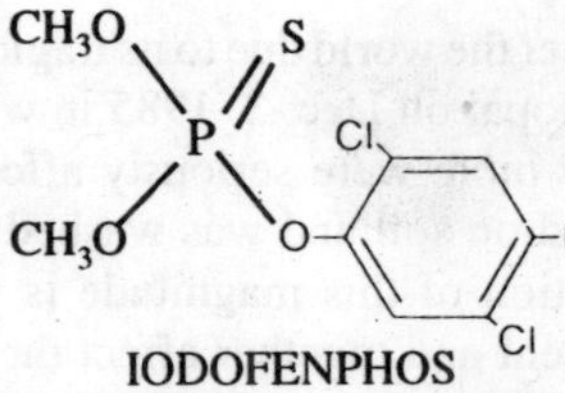

IODOFENPHOS
O, O -dimethyl - o - 2.5 dichloro - 4 - iodophenyl phosphorothioate

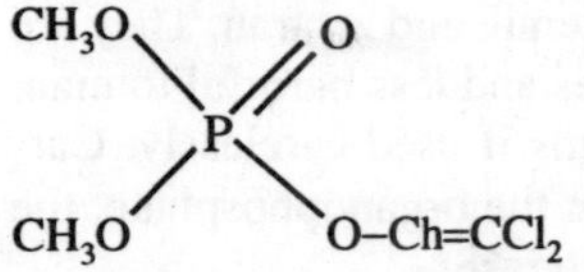

DICHLORVOS
2.2-dichlorovinyl dimethyl phosphate

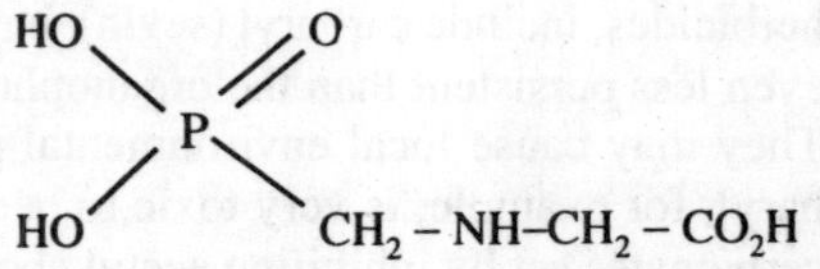

GLYPHOSATE

Fig. 6.7

Malathion (Fig. 6.7) is less toxic to mammals than Parathion. Consequently it is used for the protection of a wide range of crops including cotton, fruit, potatoes, rice, vegetables, stored grain and mammalian parasites.

Chlorpyrifos was introduced in 1965 for use in major crops such as citrus fruits coffee cotton maize and also against the mosquito. The dose rate is in the range of 200-1200 g/ha. In soil it persists for only 60-120 days, being cleaved by hydrolysis into phosphate and trichloropyridine-2-ol.

Iodofenphos, is also a contact insecticide, like the above compounds, but in addition it has systemic properties. That is to say it is translocated in plants and persists long enough to make the host itself toxic to insects that eat or suck it, but not to those that merely alight on it. It has an important application in poultry houses.

Dichlorvos is of low molecular weight and is very volatile, Due to this it is incorporated into slow release insecticidal strips for control of flies, moths and mosquitoes. It is also effective in plants against sucking insects and as an anthelmintic. Although appreciably toxic to mammals it is not persistent being rapidly hydrolysed.

Carbamates

Among the carbamate pesticides, methyl isocyanate or MIC (CH_3NCO) as an environmental pollutant has drawn attention of public all over the world due to its tragic leakage from the Union Carbide Factory in Bhopal on Dec. 3, 1985 in which over ten thousand persons died and many more were seriously affected. Much of the leaked gas ultimately settled on soil and was washed out to water bodies. Although pesticide pollution of this magnitude is very rare yet smaller incidents are very frequent and together affect the health of about a million persons every year all over the world. The synthesis of MIC requires the use of phosgene ($COCl_2$) which is an extremely poisonous chemical.

Carbamate insecticides, also used as molluscicides, fungicides and herbicides, include carbaryl (sevin), baygon, temik and zectran. They are even less persistent than the organophosphates and less harmful to man. They may cause local environmental problems if used carelessly. Carbaryl, for example, is very toxic to bees. Like the organophosphate, the carbamates act by inhibiting acetyl choline esterase.

Carbaryl is a cheap and widely used compound. (Fig. 6.8) Introduced as early as 1956 by the Union Carbide Company, it is synthesized by the reaction of α-naphthol with phosgene, which is essentially a dibasic acid chloride.

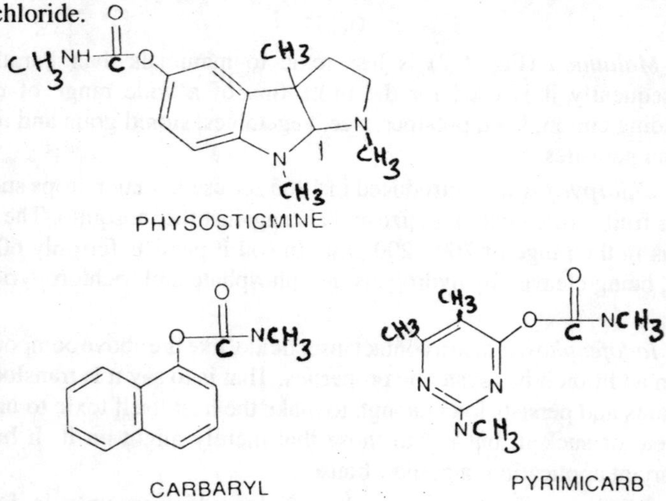

PHYSOSTIGMINE

CARBARYL

PYRIMICARB

Fig. 6.8 Carbamate pesticides.

The mammalian toxicity is moderate; for the rat the LD_{50} is 500 mg/kg while 200 μg/g in the diet showed no effect over 2 years.

Pirimicarb was introduced in 1968 by ICI after it was discovered that carbamates which had a pyrimidine carrier group were less toxic to mammals. It is more expensive than carbaryl but is effective against aphids including those which have become resistant to organophosphorus insecticides.

Naturally Occurring Pesticides

Pyrethroids

Pyrethroids are the insecticidal constituents of pyrethrum flowers (*Chrysanthemum cineriaefolium* and Chrysanthemum *coccineum*). Five insecticidal compounds, pyrethrins I and II, cinerins I and II, and jasmolin II, are present in the achenes of the flowers from which they may be extracted with organic solvents. These naturally occurring pyrethroids are unstable to the action of sunlight, air, moisture and alkalis, and are rapidly degrated after application.

Pyrethroids have a low toxicity to mammals because they are rapidly metabolized to harmless substances. However, they can cause severe allergic dermatitis and systemic allergic reactions. Large amounts may cause nausea, vomiting, headache and other disturbances of the central nervous system.

The insecticidal activity of the flowers was known to the Chinese in the first century AD but commercial production of pyrethrum did not begin until 1850. About half the current world production comes from Kenya; other major sources are Russia and Japan.

Each flower head contains 2—4 mg (≃ 2%) of a mixture of active components and these are extracted with light petroleum. The four

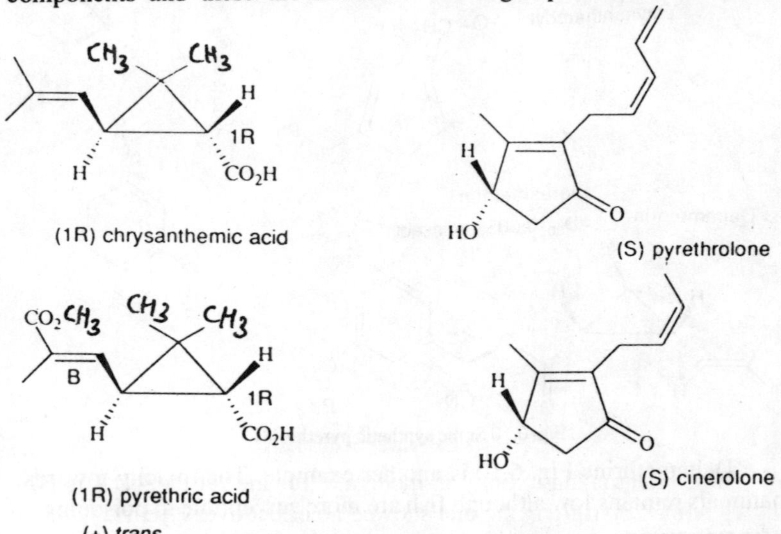

CH_3 CH_3 H 1R H CO_2H

(1R) chrysanthemic acid

H HO O

(S) pyrethrolone

CO_2CH_3 CH_3 CH_3 H B 1R H CO_2H

(1R) pyrethric acid

(+) *trans*

H HO O

(S) cinerolone

Fig. 6.9 Principal components of the pyrethrins.

principal compounds are esters formed by combination of either of the acids, chrysanthemic Fig. 6.1 and pyrethric Fig. 6.1, with one or other of the alcohols, pyrethrolone and cinerolone Fig. 6.1.

Synthetic analogues of pyrethrin. The high cost of extraction of natural pyrethroids has led to the production of similar synthetic compounds, of which the best known are the allethrins. The synthetic compounds have a wide range of properties and it is impossible to generalize about them. The first of these compounds was allethrin Fig. 6.10 prepared from a racemic mixture of (*R,S*) chrysanthemic acids. In bioresmethrin Fig. 6.10 the cyclopentenone ring of the natural product is replaced by the structurally simpler benzylfuran and yet the compound is extremely potent, although still short-lived due to its photosensitivity. Some of the thinking behind the commercial development is illustrated by permethrin Fig. 6.10, which includes the easily accessible phenyl ether group and the inclusion of two chlorine atoms in the chrysanthemic acid part.

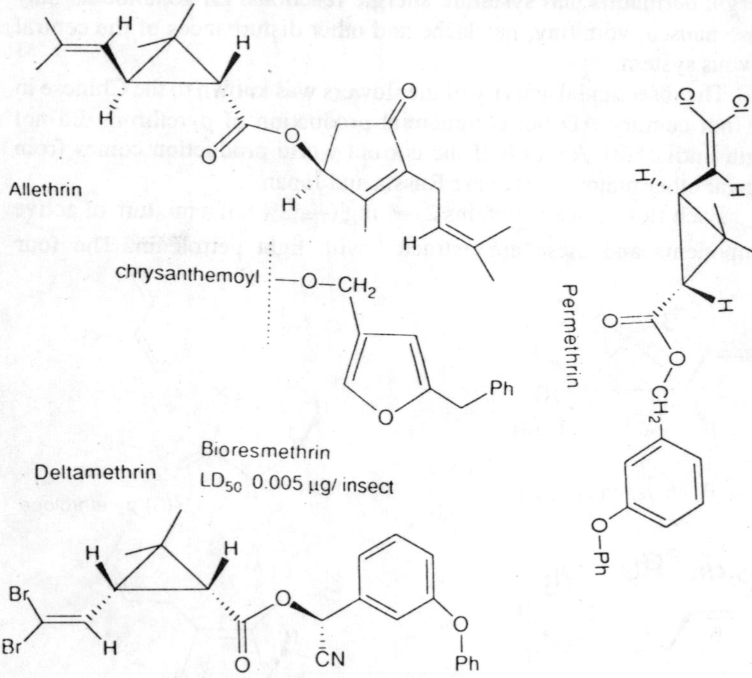

Allethrin

chrysanthemoyl —O—CH_2

Bioresmethrin

LD_{50} 0.005 μg/ insect

Permethrin

Deltamethrin

Fig. 6.10 Some synthetic pyrethroids.

Deltamethrine Fig. 6.10 is another example. The toxicity towards mammals remains low although fish are more susceptible to poisoning.

HERBICIDES

A class of chemicals commonly called "weed-killers" that will kill plants or otherwise interfere with their growth. Some of the more common herbicides are included in Table 6.1.

Herbicides can be non-selective (i.e., kill all types of plants commonly used to clear garden paths of weeds) or selective (i.e. can be applied to a crop in order to kill unwanted weeds, without damaging the crop itself). Snell (1986) identifies two types of herbicides: those that work by foliar application (Contact herbicides) and those that are applied to the soil (Soil Sterilants). The former are divided into two groups: seven systemic groups, which include **2,4,5-T paraquat** and **glyphosate** and four contact herbicides, including **ioxynil, dinoseb**, and **pentanochlor**.

There are nine groups of soil application herbicides, including **diuron, atrazine** and **nitrofen**.

A selective herbicide is one that acts only against certain types of plants and not all types (e.g. 2,4-D kills broadleaf weeds in lawns with little or no damage to lawn grass). A herbicide is described as preemergent if it should be applied before weed seeds have germinated and the weed plants have energed from the soil. Other terms that refer to the time at which a herbicide should be applied are preplant (*i.e.*, application of herbicide should be done before the land is seeded with the desired crop), postemergent, and postharvest.

Soil Sterilants

Soil sterilants are compounds that are mixed with the soil before planting, to kill plants and animal pests in the range of the substance. Methyl bromide is an example of this type. Soil sterilants soon diffuse out into the atmosphere and soil organisms eventually return to the soil.

Table 6.1 Some Common Herbicides

Name	*Structure*	*Comments*
2,4-D (2,4-Dichlorophenoxy acetic acid)	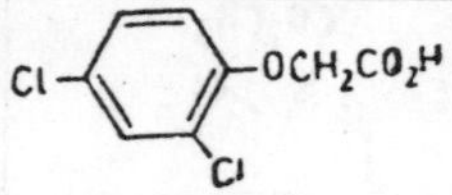 Cl, OCH_2CO_2H, Cl	Used as a salt or easter; Selective; persists in soil for 2-3 weeks. Heavily used in Vietnam war.
2,4,5-T (2,4,5-Trichloro-phenoxy acetic acid)	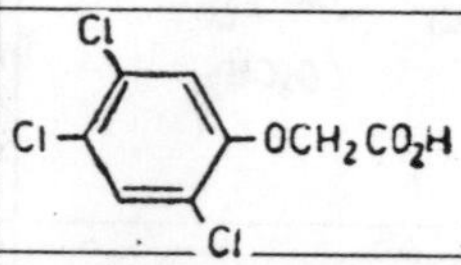 Cl, Cl, OCH_2CO_2H, Cl	Used as salt or ester.
Silvex (2,4,5-TP)	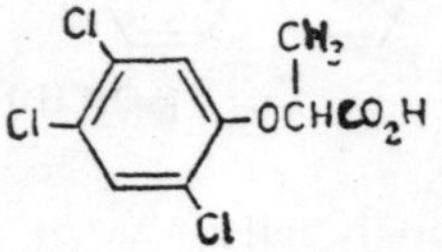 Cl, Cl, CH_3, $OCHCO_2H$, Cl	Used as salts and esters. Mostly for control of woody plants and aquatic weeds but good (in combination with 2,4-D) against most weeds in established lawns. Activity about equal to 2,4,5,-T on many woody plants, including oaks.

Name	*Structure*	*Comments*
PCP (Pentachlorophenol)	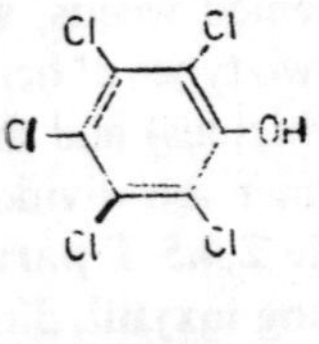	Some preemergent and some postemergent with feed crops, cotton; also a defoliant, fungicide, insecticide and wood preservative; suspected of affecting DNA; very toxic to man and animals.
Picloram (4-Amino-3,5,6-trichloro pi colinic acid)	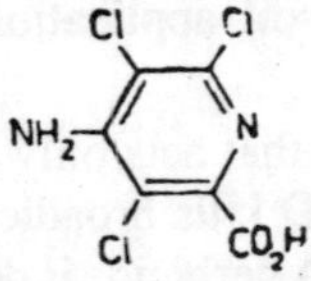	Used as salts; extremely persistent—perhaps the most persistent and most active herbicideknown; will killtrees if applied to bark at the base.
Dicamba (Banvel-D)	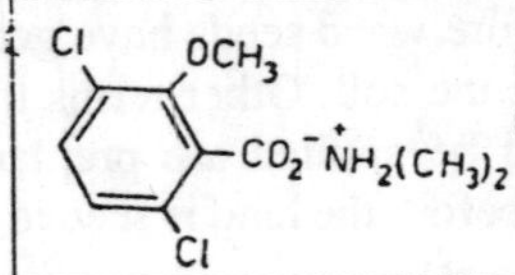	Selective; mostly postemergent; good against annual broadleaf weeds and some perennial weeds; used on golf courses and noncrop areas.
Dichlorprop [2-(2,4-Dichloro- phenyoxy) propionic acid]; 2,4-DP	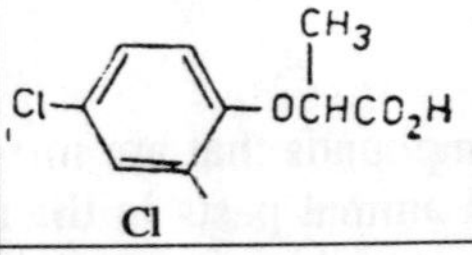	Against brush in rangeland clearance; on established lawns to kill emerged, broadleaf weeds.
Mecoprop [2-(2-Methyl-4-chlorophenoxy) propionic acid; MCPP]	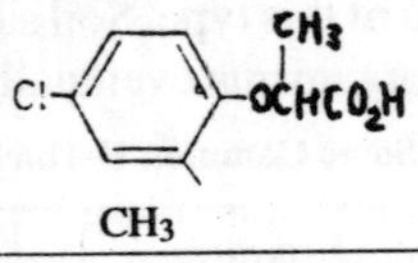	Selective; postemergent; found in some home lawn fertilizers; persists in soil several weeks; safer than 2,4-D on sensitive turf.
DCPA (Dimethyl terrachloroterephthlate; Dacthal)	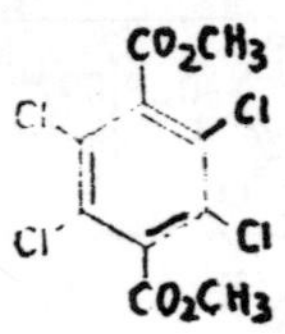	Selective; mostly preemergent; applied at time of seeding of alfalfa, vegetables, cotton; probably most versatile for home use: good against seeds of crabgrass and purslane; does not kill emerged broadleaf weeds on lawns.
Paraquat	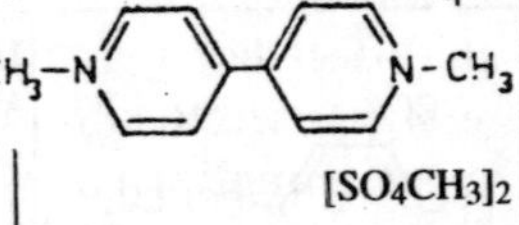 $[SO_4CH_3]_2$	Nonselective; very toxic; contact; noncrop weed control; postemergent; known to induce large chromosome alterations (but not point mutations) in test species.

Name	*Structure*	*Comments*
Atrazine (AA trex)	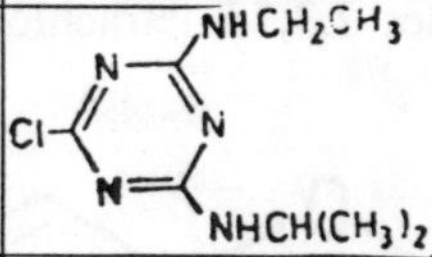 Cl-triazine ring with $NHCH_2CH_3$ and $NHCH(CH_3)_2$	Selective; mostly preemergent or postharvest; used for wheat, sugarcane, and sorghum.
Diuron	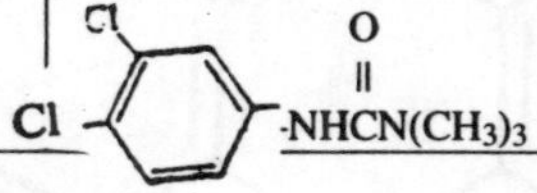 Cl, Cl-phenyl –$NHC(=O)N(CH_3)_3$	One of the best for longterm vegetation control.

Contact Herbicides

The contact herbicides are quick-acting substances that kill plants by direct contact with their leaves. Examples are DNOC (dinitrocresol) and PCP (pentachlorophenol). They are not effective against perennials with roots that can form new shoots. DNOC is quite toxic, and tends to accumulate in an animal system. It can be absorbed in toxic amounts either directly through the skin or through the food chain.

Systemic Herbicides

Systemic herbicides diffuse through leaves or roots into the plants system for distributing its fluids. This group includes more selective herbicides. One such group of compounds have hormone-initiating action. Herbicides such as **2,4-D, 2,4,5-T**, **silvex, dicamba**, **mecoprop** and **picloram** mimic the plants own hormones but not exactly in the manner in which the plant needs. Inspite of much research to determine how they kill only certain weeds, their mechanism of action is generally unknown. Part of their selectivity against broad-leaf results from greater absorption and translocation than in grasses, but more important factors are involved. It is sometimes stated that they cause a plant to "grow itself to death", but this is misleading. Modern hypotheses suggest that these compounds alter DNA transcription and RNA translation, so that the proper enzymes needed for coordinated growth are not produced properly, but how this is accomplished is unknown.

2,4-D and 2,4,5-T are probably the most widely used herbicides. They are relatively cheap and affect dicots more than monocots. Because of their selectivity they are often used to kill broad-leaf dicot weeds in cereal grains and lawns. **Agent orange**, which was used in the Vietnam war as a defoliant, is an effective mixture of 2,4-D and 2,4,5-T. A common popular misconception seems to be that, 2,4,5-T is a very toxic substance to animals, whereas there is overwhelming scientific evidence that the pure compound is of low toxicity compared with many chemicals in daily use in agriculture.

The toxicity of commercial samples of 2,4,5-T is due to its contamination with a by-product 2,3,7,8-tetrachlorodibenzo-*para*-dioxin (TCDD).

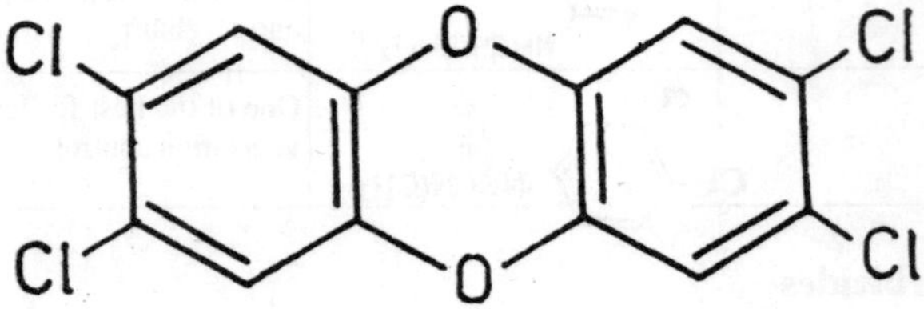

TCDD

TCDD is an extremely toxic substance and the concern over 2,4,5-T has arisen from the use of TCDD-contaminated samples in some areas. Normal application rates of 2,4,5-T approximate 1 kg/acre, and commercially available formulation must contain less than 0.1 ppm of TCDD. Thus, normal application should disperse a maximum of 1 ppm TCDD per acre.

NEMATICIDES

Against Plant Parasitic Nematodes

The nematicides can affect large areas of land. The main nematicides used are 1, 3-dichloropropene (telone) and 1, 2-dibromoethane, both of which are liquids applied directly to the soil. Because they are phytotoxic they can only be used in the absence of growing crops. These compounds are also very toxic to mammals. They irritate the skin, eyes and mucous membranes, damage the liver, and ultimately cause death from severe lung injury. Other soil nematicides in use are sodium N-methyldithiocarbamate in aqueous solution 0-2, 4-dichlor-phenyl-O-diethylphosphorothioate (V-C 13 nemacide) as an emulsion, calcium cyanamide, urea, and 3, 5-dimethyl-tetrahydro-1,3,5 2H-thiadiazine-2-thione (mylone). The last three chemicals are solids and must be thoroughly mixed with the soil to be effective. All of these compounds are toxic to mammals and probably to most other animals as well.

Fumigants, such as methyl bromide, may be used in confined spaces, or under polythene sheeting, where only small areas require to be treated. Methyl bromide is very toxic and exposure of mammals to 50 mg per body weight can cause death.

Seeds and bulbs may sometimes be treated directly with dilute formaldehyde solution in water (0.5 to 1.0% weight to volume), with or without a trace of detergent. It is unlikely that this constitutes an environmental hazard.

Against Mammalian Parasitic Nematodes

Whether the drugs used in the treatment of nematode infection in human beings and domesticated animals constitute an environmental hazard does not seem to have been assessed. However, since at least 200 million people throughout the world are thought to be infected with nematodes, and since presumably a similar proportion of the domesticated animal population is at risk, the use of nematicidal drugs must be considerable. In consequence, appreciable amounts of these compounds or their derivatives must enter the environment. It is, therefore, justifiable to describe briefly here the principal drugs involved.

Diethylcarbamazine is the most effective drug against filarial nematodes. It is not highly toxic but it can nausea, vomiting and headache in humans, and muscle tremors and convulsions in dogs. Thiabendazole is the most widely used drug for the control of gastro-intestinal nematodes, other than *Trichuris* spp., in ruminants. It is also used on citrus fruit as a fungicide. Its acute toxicity to mammals is low. Phenothiazine is often used against intestinal nematodes in sheep and chickens. It is not used on cattle, horses or humans, because of its many toxic effects on these species. Piperazine is the drug of choice for the treatment of oxyuriasis and ascariasis in both humans and other mammals, since it is highly active against the nematodes responsible but almost nontoxic to mammals at the therapeutic dose levels.

FUNGICIDES

Probably the most widely used agricultural fungicides are *Bordeaux* mixture and sulphur. Bordeaux mixture is made by combining a solution of copper sulphate with a suspension of calcium hydroxide, usually in equal amounts by weight. Bordeaux mixture may have harmful side effects on treated plants, e.g., it may cause accelerated transpiration and, hence, skin russets of apples, and premature defoliation of peach trees. Other effects may be associated with localized concentrations of copper. Sulphur has low toxicity but it may cause irritation to the skin, eyes and respiratory tract. It may also be converted to hydrogen sulfide which is a very powerful poison.

Organic fungicides are replacing *Bordeaux mixture* and sulphur in many applications. Organomercurial compounds are used as seed dressings. Dithiocarbamates such as ferbam, ziram, nabam, zineb and maneb, are widely applied to vegetables and ornamental plants. Other organic fungicides in use are thiram, captan, glyodin, chloranil, dichlone, dodine (cyprex) and karathane. Of these, thiram, dodine and karathane are appreciably toxic to humans. Captan, glyodin, chloranil and dichlone have low toxicity for humans because they are poorly absorbed from the gut.

Hexachlorobenzene, which has been used as a fungicidal seed dressing, has a relatively low acute toxicity for mammals but can cause serious delayed effects. This was seen most dramatically in Turkey in 1955 when up to 5000 people were poisoned by eating hexachlorobenzene-treated grain. Their symptoms included enlarged livers, abnormal light sensitivity, weight loss and abnormal hair growth, particularly on the face. This last symptom gave the epidemic its name of monkey face disease. It is worth noting that hexachlorobenzene may be an impurity in other pesticides, at levels up to 10 per cent of the total. It is chemically stable and, therefore, persistent. Measurable amounts are released into the environment during the manufacture of perchlorethylene.

The antibiotic, cycloheximide, has been used to treat cherry leaf spot. This antibiotic is extremely toxic, causing gastro-intestinal lesions, coma and death. Sympathomimetic drugs may alleviate the symptoms and hydrocortisone or adrenal cortical extract may prevent death.

In addition to the fungicides which are applied directly to plants or seeds, there are others used to clear fungi from the soil before planting crops. Many of these, *e.g.*, formaldehyde, chloropicrin and methyl-isothiocyanate, are phytotoxic but their volatility enables them to escape from the soil before planting commences. However, these compounds are also highly toxic to animals and this should be borne in mind when their application is being planned. Pentachloronitrobenzene is a soil fungicide which appears to have low phytotoxicity and it has been applied directly to soil in which cruciferous crops and lettuce are growing. However, it is moderately toxic to animals and accumulates in body fat. Hence, there is a potential longterm risk similar to that from chlorinated hydrocarbons.

REFERENCES

Carter, L.J. 1976, Michigan's PBB incident: Chemical mix-up leads to disaster. *Science*, 192, 240-243.

De Bach, Paul. 1964. *Biological Control of Insects Pests and Weeds.* Reinhold Publishing Corporation. New York.

Gunn, D.L. and J.C.R. Stevens, (Eds). 1976 Pesticides and Human Welfare. Oxford University Press.

Higgins, I.J. and R.G. Burns, 1975. *The Chemistry and Microbiology of Pollution.* Academic Press, London.

Varo, P. (Ed). 1976. *Pesticide Chemistry*—3 3rd IUPAC Congress on Pesticide Chemistry- Butterworths. London.

CHAPTER 7

Solvent and other Organic Chemicals

Organic solvents are today produced almost entirely from petroleum. Table 7.1 lists the solvents (which are in most common use) with their principal properties and application.

Table 7.1 Properties of some important solvents

Compound	Boiling point (°C)	Odour perception	Applications
Acetic acid	118	0.1-1.5 mg/m^3	Synthesis, pharmaceutical, photographic, polymers
Acetone	56	5-50 mg/m^3	Solvent for cellulosic resins and vulcanisates, adhesives, paints
Benzene	80	3 mg/m^3	Motor fuels, synthesis especially maleicanhydride, flavours, perfumers, paints
Carbon tetrachloric	77	500 mg/m^3	Manufacture of fluorocarbons
Chloroform	62	1000 mg/m^3	Refrigerants, pharmaceuticals
Cyclothexane	81	4-10 mg/m^3	Nylone synthesis
1 4-Dioxane	101	4-25 mg/m^3	Lacquers, paints, cosmetics, cleaning and detergency
Ethanol	78	5-5- mg/m^3	Synthesis, brewing, soaps, cosmetics
Ethyl ether	35	1-10 mg/m^3	Manufacture of ethylene, perfumery, extraction
Method	65	1-100 mg/m^3	Synthesis of proteins, esters and H.CHO for plastics. Dehydration of natural gas

(*Contd.*)

Table 7.1 (*Contd.*)

Compound	Boiling point (°C)	Odour perception	Applications
Tetrahydrofuran	66	10 mg/m^3	Polyether synthesis PVC solvent
Toluene	111	1-100 mg/m^3	Fuels, paints and coatings, isocyanates, benzene substitute
Trichloroethane	74	540 mg/m^3	Metal cleaning, protection of upholstery, adhesives
p-Xylene	138	1010 mg/m^3	Synthesis of terephthalic acid, motor fuels

A detailed study of solvents encompasses a large proportion of organic chemistry and the discussion here is limited to those which offer the greatest environmental risk. These are formulated with volatile solvents so that once applied they rapidly become tacky and ready to bond. The solvents are therefore liable to escape when use domestically and in small factories with inadequate control measures.

1,1,1-Trichloroethane (TCE) is one of the most used solvents as it is non-flammable and dries rapidly. Between 40-50 thousand tonnes are used annually for adhesive production. It is dangerous when inhaled and can cause death through congestion of the lungs or from heart failure. Other solvents in common use are CH_2Cl_2, toluene, *n*-hexane, methyl ethyl ketone and ethyl acetate. Major uses are in bonding decorative laminates (*Sunmica*) to practicle board or plywood, also in bonding polyurethane foam to furniture frames.

Apart from the uses just mentioned it replaced ethyl acetate in adhesives for packaging; it was also applied in latex ahdhesive based on styrene-butandiene formulations. TCE is mainly used in metal cleaning in adhesive and in aerosols, electronics and other applications. With concern about damage to the ozone layer TCE is no longer favoured; it has an *Ozone depletion potential* (ODP) of 0.1. This substitution may protect the ozone layer but release of these solvents enhances capture of NO so further increasing tropospheric ozone levels.

Other protective measures include greater use of water-based latexes and emulsions. These are ineffective for the bonding of rubbers and some plastics; when sprayed they also produce fine mists which are carried around the workplace. Solid adhesives consisting of isocyanates and polyurethanes which cure when exposed to moisture are also available.

Solvent release may be avoided by the hot melting of plastics such as polyethylene and polyesters which bond to metals, plastics and paper on cooling. This technique is sued in bookbinding, glazing and for the installation car carpets and panels. Curing of acrylic, polyester and urethane resins by irradiation with UV, IR light or an electron beam is being used increasingly. These alternatives require expensive new equipment but are generally not more costly overall.

Solvent in Coatings and inks

In 1986 almost half of coatings were solvent based which included TCE. Its use was restricted to thinning for spraying inks on to wallpapers, and for labelling bags, bottles and cartons. Here also the use of water-based materials and radiation curing controls solvent release.

Solvent in Aerosols sprays

The much publicized concern about damage to the ozone layer arises from the use of CFCs in spray cans . These formulations include three components: (i) the active ingredient (ii) a solvent and (iii) propellent.

It has been estimated that in 1986 6.8 billion individual items were produced and many included TCE as solvent. At this time production in the USA was 18 600 tons and in Europe 12 400. In Japan production then at 10 800 tons had fallen to 5000 by 1990. Alternative solvents are dimethoxyethane and low boiling alcohols, ketones and petrol fractions.

Solvents in Metal cleaning

At present CFCs are used for cleaning and drying a wide range of parts as also are per chlorethylene and methylene chloride. Parts which require cleaning in solvent baths include Car engine parts, adhesive spreaders, silk screen stencils, polymer formers, oil rig equipment, photocopiers and printing machine.

Nissan (Japan) will replace CFC-11 (CCl_3F), CFC-12 (CCl_2F_2) and CFC-113 (CCl_2F-$CClF_2$) with HCFC-134 (CH_3-CH_2F) by the end of 1994.

Dry cleaning of clothes

Organic solvents are used as, unlike water, they do not distort the fibres. The machines have long been totally enclosed on account of cost control. The solvents available are listed in Table 7.2. According to the provisions of the Montreal Protocol for protection of the ozone layer, CFC-113 (ODP 0.8) and TCE (ODP 0.1) will not be acceptable and should be phased out by 1977. This presents practical problems since dry cleaning equipment has a lifetime of about 15 years and is only suitable for the solvent originally specified. Even HCFC-225, the recommended substantiate, will be unacceptable by the year 2020.

Carbon tetrachloride is still used for dry cleaning in Eastern Europe, the former USSR and in South East Asia. It is produced by the high temperature chlorination of methane, propylene or carbon disulphide for

Table 7.2 Some solvents user in dry cleaning

Solvent	*Boiling point (°C)*
CFC-113	48
Perchloroethylene	121
Petrol based	150-210
1, 1, 1-TCE	74
HCFC-225 ($CF_3CF_2CHCl_2$)	53

use as a feedstock in the synthesis of CFC-11 (CCl_3F) and CFC-12 (CCl_2F_2). As these solvents are phased out it will only be used in a limited number of special syntheses, for example that of picloram.

Toxic Effects of Solvents

The occurrence and toxicology of solvents in water has been extensively studied in the United Kingdom by the Water Research Centre based at Medmenham. Apart from protecting potable water supplies the analysis reveals those substances released by industry whether by evaporating to air or by direct discharge to rivers as waste.

Aliphatic hydrocarbons. Some 60 compounds in this group have been detected in air and water consistent with their use in fuels.

In general they do not present much risk to mammals but butane and n-hexane have attracted attention.

Butane accounts for 30% of all solvent abuse, *n*-Hexane is one of the isomers present in motor fuel and in alkane mixtures used for paint thinning and for metal cleaning. It forms 1% of diesel emissions and 1.2% of the exhaust from petrol engines while losses from fuel tanks and carburettors can reach 10%. *n*-Hexane has a lifetime of about 6 hours in smog.

On exposure for 10 minutes to levels of 2000 ppm there was no acute response in man. The principal chronic effect in mammals and man from exposure is one of neurotoxicity.

Abnormalities in the nervous system have been noted in workers occupationally exposed. *n*-Hexane is rapidly metabolized in mammals being excreted as the alcohol,

$$CH_3{-}\underset{\displaystyle OH}{\underset{|}{CH}}{-}CH_2{-}CH_2{-}\underset{\displaystyle OH}{\underset{|}{CH}}{-}CH_3$$

Its neutroxicity is attributed to further oxidative metabolism to form 2,5-hexanedione.

$$CH_3{-}\overset{\displaystyle O}{\overset{\|}{C}}{-}CH_2{-}CH_2{-}\overset{\displaystyle O}{\overset{\|}{C}}{-}CH_3$$

Aromatic hydrocarbons. On inhalation acute effects include giddiness, headaches and nausea; chronic poisoning leads to depression of bone marrow function. There is evidence of benzene being a tumor promoter as it is linked to myeloid leukemia and lymphomas. In the USA, benzene has been banned as a laboratory solvent but it still forms up to 5% (vol.) of lead-free petrol. In contributes to exhaust emissions (2.4%, vol. of total hydrocarbons). The maximum allowable levels for an 8 hour day and 40 hour day and 40 hour week range between 5 and 10 ppm.

Toulene is now preferred to benzene as a solvent. Death due to its inhalation has occurred as a result of solvent abuse and at high levels of occupational exposure (30 ppm) it can induce mild abnormalities of the CNS. Levels in petrol and air are similar to those of benzene' its lifetime in smog is about 6 hours.

Organochlorine compounds

Vinyl chloride. This compound presents the most serious risk although it is still used in massive almost for the production of PVC. Its B.P. of –14°C makes for a problem of containment. Vinyl chloride causes liver degeneration in mammals and is carcinogenic in man, including tumors of the live and blood in those occupationally exposed; it is suspected as a cause of human mutations. Accepted levels in the workplace are in the range 3-5 ppm.

Chloroform. This occurs in air at levels up to 5 $\mu g/m^3$ and has been detected in foodstuffs 30 mg/kg. It is common in water samples being produced by the haloform reaction during chloination of water and sewage; in air it may be formed by the photochemical degradation of trichloroethylene.

Chloroform causes kindly damage and is cancer inducing in rats and mice. Studies showed that chlorination of tap water did not lead to enhanced chloroform levels in human blood plasma; the WHO has set a guideline of 0.03 ppm for chloroform in drinking water.

Carbon tetrachloride. This compound is on the decline as an industrial chemical. Acute exposer in animals produces death by depression of the CNS or through liver damage. In the liver it is metabolized via the trichloromethyl radical (Fig. 7.1) producing chloroform, dichlorocarbone, and phosgene, which is also a metabolite of chloroform. Capture of oxygen yields the long-lived peroxy radial which is held responsible for liver damage.

Carbon tetrachloride causes liver in animals and there is a possibility that this extends to humans.

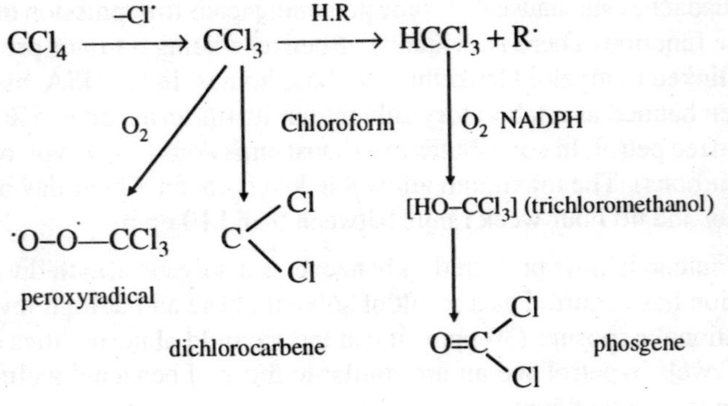

Fig. 7.1 Metabolism of carbon tetrachloride.

1, 1, 1-Trichloroethane. This an important industrial solvent . It occurs in air at about ' µg/g and has been found in food and water samples. It does not appear to be metabolized as CCl_4, being largely exhaled from the lungs unchanged.

Deaths from acute occupational exposure have been recorded and solvent abusers are also at risk of heart failure. There is no evidence of long-term chloric effects but the accepted workplace level is 10 ppm.

DETERGENTS

Detergents are a group of synthetic compounds within the general class of surfactants. They were first produced during the First World War in Germany. The period after the Second World War saw a dramatic escalation in their use. The principal products were based on cheap and readily available petrochemicals.

A typical detergent is shown in Fig. 7.2.

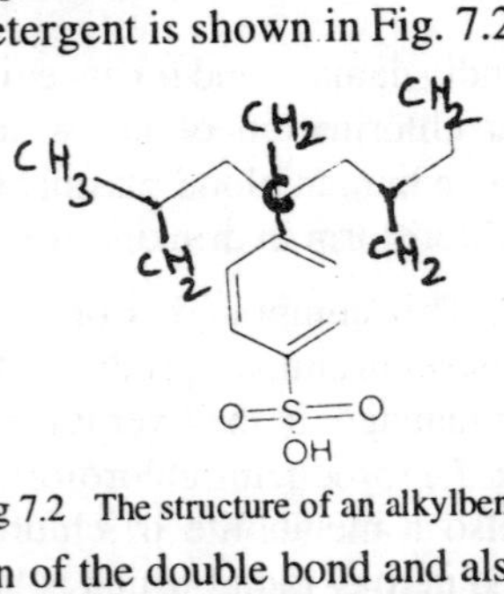

Fig 7.2 The structure of an alkylbenzene sulphonate.

The position of the double bond and also the number of polymerized alkene molecules varies because it is not economic to separate single substances, nor is a mixture of alkyl groups any less effective for detergency.

Types of detergent. This sulphonate (Fig 7.2), like soaps of the stearate group Fig. 7.4 dissolves in water and becomes effective as its anion. These compounds are therefore known as anionic detergents. The sulphonates have the advantage that, unlike the stearates, they are not precipitated in hard water as calcium or magnesium salts. Booth types depends for their action on removing organic material from sold surfaces on the residence of the hydrophilic head in the aqueous and the lipophilic tail in the organic phase (Fig. 7.3). This greatly reduces the surface tension allowing the non-aqueous materials to be wetted and washed out. Above a certain critical concentration the surfactant will support water-insolubles within aggregates or micelles, which present the heads to the water while the tails from an internal clump.

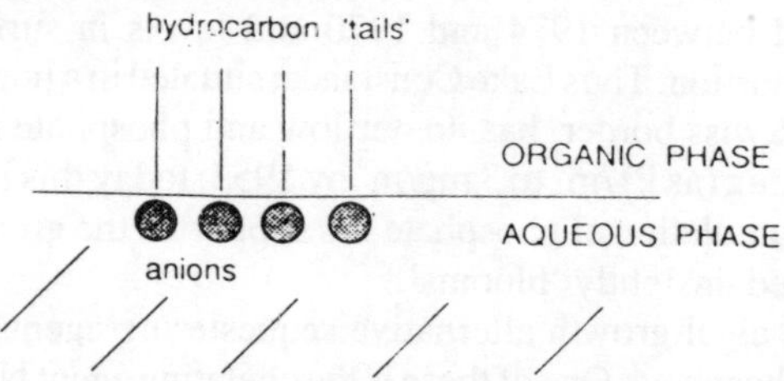

Fig. 7.3 Phase interaction of an anionic surfactant.

Anionic detergent still occupy a dominant place in the market, but the use of detergents is widespread in industry and the home. This necessitates the formulation of many specialist types including cationic and non-ionic detergents (Fig. 7.4).

$CH_3—(CH_2)_{16}—C(=O)O^-Na^+$ A soap - sodium stearate

Cationic detergent

$CH_3–(CH_2)_{14}–CH_2—N^+(CH_3)_3\ Cl^-$ Hexadecyl trimethylammonium chloride used for home laundry and fabric

non-ionie detergent

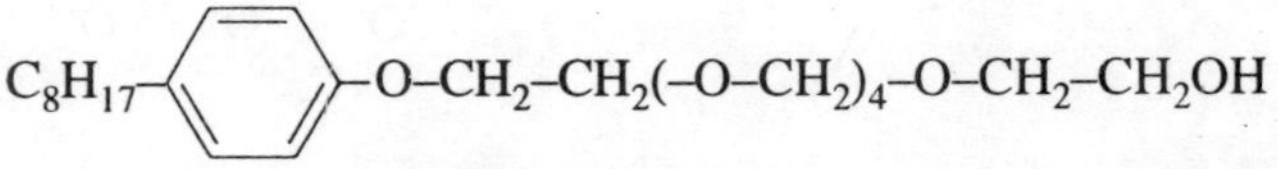

$C_8H_{17}–C_6H_4–O–CH_2–CH_2(–O–CH_2)_4–O–CH_2–CH_2OH$

hexaethoxyethylene glycol monoether of p-octyl phenol

Fig. 7.4 Other detergents

Builders. In the original preparation of sulphonates, sulphuric acid in the residues was neutralized so as to bulk up the product and make it easier to measure out. This practice continues and in addition other builders are added, including sodium carboxymethylecellulose to prevent redeposition of dirt and phosphates to remove hardness from the water.

Pollution by detergents. The increase in the use of anionic sulphonates based on propylene led to concern about persistence and foaming when they were released to water bodies. The traditional soaps are degradable by the effluent bacteria but the branched chain of the sulphonate are resistant. Consequently by the early 1960s industry began to formulate these detergents from degradable straight chain polymers of ethylene.

The production of sodium tripolyphophate as an additive increased one hundredfold between 1974 and 1970 and levels in surface waters increased in proportion. Thus Lake Constance, situated in a populated area on the German-Swiss border, has no outflow and phosphate rose from a natural level 0.2 mg (as P) /m^3 to 3 mg/m^3 by 1953: today this is very high. This type of accumulation of phosphate has supported the growth of large and the associated unsightly 'blooms'.

To control algal growth alternative sequestering agents have been substituted for phosphate. One of these is the chelating agent Nitrukiacetic acid, which associates with calcium and magnesium as shown in Fig. 7.5. Nitriloacetic acid (NTA) is, however, environmentally suspect since is also sequesters and therefore mobilizes heavy metals such as lead and cadmium from sediments. The alternative to phosphate are zeolites which are polymeric sodium aluminium silicates. A simplified form is the structure shown in Fig. 7.6, where the trivalent aluminium atoms in sharing with the tertravalent silicon acquire a surplus negative charge; as charge; as a result the zeolite associates with calcium ions in order to achieve electrical neutrality.

```
                    O
O          CH2–C//
 \\        /     \O
  C—CH2—N ....... Ca++
 /         \      O
O           CH2–C/
                \\
                 O
```

Fig. 7.5 association of sodium nitriloacetic acid with calcium ion

```
     Ca··          Ca··
   O·     O·     O·
   |      |      |
–O–Al–O–Si–O–Al–O–
   |      |      |
   O·     O·     O·
     Ca··          Ca··
```

Fig. 7.6. simplified structure of a zeolite

REFERENCES

Davidsohn, A.S. and Milwidsky, B. (1987) *Synthesis Detergents* (7th edition), Longman, Harlow.

Elsom, D. (1987) *Atmospheric Pollution,* Blackwell, Oxford.

Goring, C. A. I. and Hamaker, J. W. (1972) *Organic Chemicals in the Soil Environment,* Dekker, New York

Harrison, R. (ed) (1990) *Pollution Causes, Effects and Control,* Royal Society of Chemistry, Cambridge.

Harrison, R. (ed) (1992) *Understanding Our Environment*, Royal Society of Chemistry, Cambridge.

CHAPTER 8

Petroleum and Related Compounds

Petroleum

Introduction

Petroleum may be toxicant in the sense that it causes harm to living organisms as a result of some chemical action. Much of the environmental damage it causes is due to its physical nature and its impermeability to oxygen. Thus, when it coats organisms, they die from asphyxiation (lack of oxygen). Similarly, contaminated areas of the environment suffer from oxygen depletion. The hydrophobic properties are associated with chemical stability. Therefore, environmental damage may occur long after input to the environment has ceased. For this reason special precautions must be taken to minimize the anthropogenic contribution.

Petroleum, generally referred to simply as oil, is a mixture of alkanes containing, on average, about 1% by weigh of aromatic hydrocarbons. It has become a matter of serious environmental concern because its extraction and use as an energy source by humans has led to its widespread distribution in the biosphere. This has been most obvious following large scale marine spillage, either from tanker wrecks or from oil-well blowouts, but the cumulative effects of the many minute spills that have occurred both onland and sea may be at least as serious.

Historical

Petroleum has been known in years BC to occur in surface seepages and was first obtained in pre-Christian times by the Chinese. The modern industry had its beginnings in Romania and in wells sunk in Pennsylvania in 1859. The principal early use was to replace expensive whale oil for lighting, but today it consumption as a fuel and its dominance of the market for chemicals has led to a daily worldwise consumption of about 65 million barrels or 3 Gt a year.

Formation Petroleum is largely formed biogenetically from matter deposited in shallow seas and subsequently compressed by the overburden of deposited clays and shales. An intermediate coal-like material formed by bacterial action on the deposits is known as kerogen. This may be one of three types formed from (a) algae, (b) marine plankton or (c) higher plants. Hydrogen sulphide, a typical product or organic decay, may also be formed.

Properties The major compounds are saturated alkanes; alkenes are absent but aromatic hydrocarbons may occur to a significant extent, for example in Borneo deposits. Other organic components of petroleum, including some which contain nitrogen and sulphur, are shown in Table 8.1

Table 8.1 Some organic components of petroleum.

1. **Alphatic compounds**

Methane	CH_4	through the range of straight and branched chains
Ethane	C_2H_6	up to $C_{76}H_{154}$
Propane	C_3H_6	

2. **Alicyclic aromatic components** +

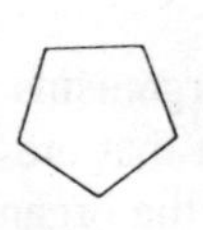

benzene, toluene, phenol, stearic acid

cyclopropane cyclopentane

3. **Nitrogen compounds**

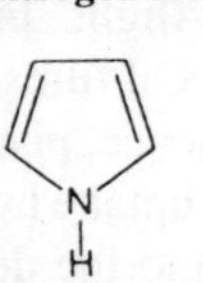

pyrrole

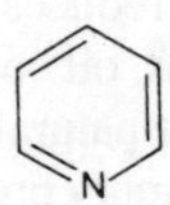

pyidine

4. **Sulphur compounds**

thiols R-SH Thioethers R-S-R; Ph-S-Ph and aromatic including:

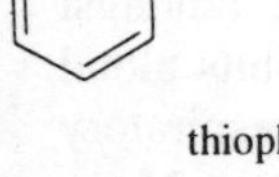

thiophenol 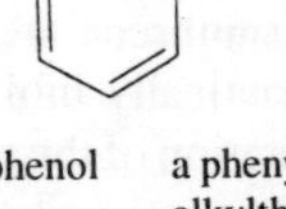 a phenyl alkylthioether 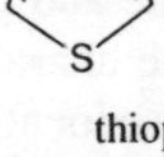 thiophene 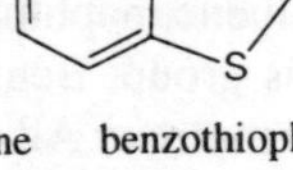 benzothiophene 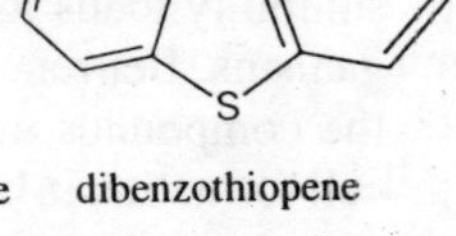 dibenzothiopene

The boiling points of some typical substances are given in Table 8.2.

Table 8.2 Boiling points of typical components of petroleum(°C)

Methane	-161	Benzene	80
Propane	-42	Toluene	111
n-Butane	-0.5	*p*-Xylene	138
n-Hexane	69	Naphthalene	218 (m.p.80)
n-Octane	126	Anthracene	340 (m.p.216)

Petroleum deposits are complex mixtures which may include as many as 40 individual substances which are not economically separable by distillation. Hence commercial practice is to sub-divide within a given boiling range (°C):

Natural gas: methane, propane, butanes
Light naphtha: 20—100
Heavy naphtha: 100—150
Kerosene:150—235
Light gas oil: 235—345
Heavy gas oil: 345-565 (steam assisted)

Toxic Effects

The harmful effects of oil on living organisms may be divided into two (i) that are primarily physical and (ii) that are primarily chemical. Physical effects are caused by oil coating the organisms or their immediate environment. This is very cleary seen when water birds become covered with oil. By matting the feathers, the oil destroys their insulative capacity reduces buoyancy in the water and prevents flight. In other organisms, oil coating may cause death by asphyxiation. Oil films on the surface or natural water reduce light transmission and, hence, photosynthetic primary production. Such films also retard oxygen uptake by water and so cause a lower dissolved oxygen concentration and the death of many organisms.

Chemical effects of oil can be related to the components involved. Low boiling point saturated hydrocarbons, at least up to octane, can produce anaesthesia and narcosis in many lower animals. Low boiling point aromatic hydrocarbons are even more toxic, and their greater water solubility tends to enhance their distribution and uptake by aquatic organisms. Benzene, toluene, naphthalene and phenanthrene are amongst the compounds in this group. Benzene characteristically inhibits blood cell formation in bone marrow. All cause local irritation of the respiratory system and excitation or depression of the central nervous system. Many may be mutagenic, carcinogenic or teratogenic. Polycyclic aromatic hydrocarbons are especially dangerous in this respect. High boiling point saturated and aromatic hydrocarbons may not exert much direct toxicity but may interfere with the responses of aquatic organisms to chemical stimuli, such as, sex attractants, with equally serious consequences.

Since many of the components of oil are chemically stable and not readily metabolized or excreted once absorbed, they are subject to food chain amplification. In this way, they may give unpleasant flavours to food for human consumption or render it toxic. It is even conceivable that oil in natural waters may concentrate fat soluble toxicants by a partitioning process, thus enhancing their toxicity.

Not only are there hazards associated directly with oil spills, there are others associated with oil clearance from natural waters and from beaches. Containment with booms followed by removal by skimming devices causes the least damage but is frequently impossible. As an alternative, detergents may be used to disperse the oil. This may give aesthetically satisfactory results but may enhance the destruction done by the spill. All detergents have appreciable toxicity and can facilitate uptake of oil by aquatic organisms. Further, they may cause oil on beaches to penetrate the sand more deeply, thereby prolonging harmful effects on intertidal animals and plants.

Polycyclic Aromatic Hydrocarbons (PAH)

The initial fusion of two benzene rings can give rise to only one product-naphthalene-but attachment of a third ring can give the linear homologue anthracene, or the branced isomer phenanthrene (Fig 8.1). These substance occur in considerable amounts in natural fuel deposits and are formed during coking of coal and on pyrolysis or fossil fuels. They are not included in the definition of a PAH which is reserved for higher homologues containing four or more benezene rings.

The number of individual PAH is considerable, bearing in mind the permutations of the modes of ring fusion and also the various positions available for insertion of side chains Fig 8.1 includes only a limited selection which includes those of commonest occurrence and/or highest toxicity.

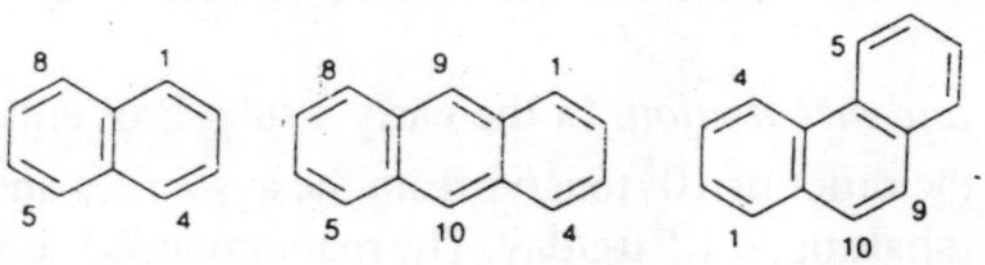

8 1 5 4 8 9 1 5 10 4 5 4 1 10 9

naphthalene anthracene phenanthrene

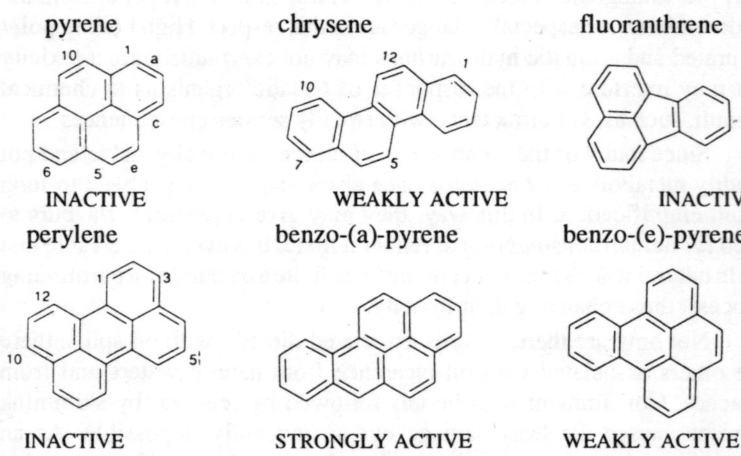

Fig 8.1 Structures of some benzene homologues and environmentally important PAH.

Toxicity. The level of activity is given alongside the structures in Figure 8.1. Benzo-[a]-pyrene (BaP) is regarded as the most dangerous of the group as it is widely distributed and strongly carcinogenic. In a test on the skin of mice, activity was noted at a does level of 5.6×10^{-5} mmol; liver damage and teratogenicity were also observed in mice.

Four fused benzene rings are a necessary but not sufficient condition for activity and a subtle structural factor is revealed since benzo-[e]-pyrene is much less active than its [a] isomer. It is known that initiation requires the binding of the PAH to cellular macromolecules DNA, RNA or protein. A key step in the activation of benzo-[a]-pyrene is its conversion into the epoxide

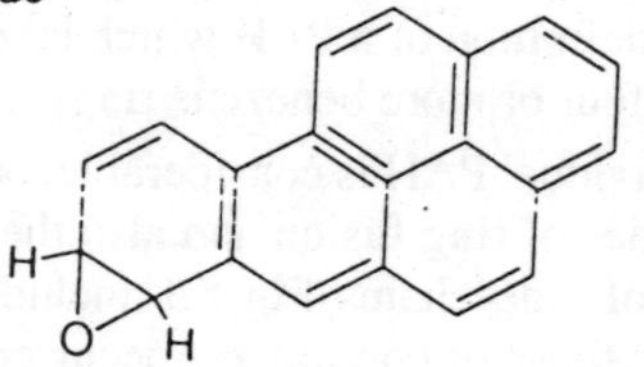

and that the ease of oxidatioin of PAHs by free radicals is indicative of their level of activity. Oxidized derivatives of PAHs also occur in the environment.

Sources and distribution. In the early 1980s BaP emissions in the USA were of the order of 10^3 tons/yr from these sources and individuals were likely to inhale up to 1.5 μg/day. The recommended acceptable level is 0.2 μ g/m^3 which was always exceeded for coke oven workers. After 5 years of exposure their risk of infection was increased ten-fold.

PAH are found at low levels of ng/g in mineral oils and in paraffin wax. They are formed by pyrolysis in petrol and light diesel engines owing to the limited air supply. If a catalytic converter is not fitted BaP is emitted at levels up to 50 μg/km travelled; the converter reduces this to a range of 0.05-0.3 μg/km. Aircraft engines typically emit 10 mg of BaP during each minute of operation. The detection of the nitropyrene is of interest and a number of other nitroderivatives have been isolated. They originate from contact with nitrogen fixation products within the engine.

PAHs enter the water body from fall-out from the above sources and also from run-off from bitumen treated roadways. The worldwide use of this material is estimate at 61×10^6 tonnes/yr with 2×10^6 tonnes applied annually in the UK and ten times this amount in the USA. The levels in domestic sewage vary but rise markedly during periods of heavy rain, for example pyrene then rises to 16×10^3 and BaP to 1800 ng/1.

PAH are found in food either by deposition on leaves (lettuce) or from heated fats (steak, sausage).

Within the house occupants are also exposed to risk from emissions from heating, for example gas central heating in a non-smoking household generates BaP in the range 0.1-0.6 ng/m^3. For smokers these limits lie between 0.4 and 1.8 ng/m^3.

The risk to smokers is of special concern and a detailed analysis of the smoke has been made. 45 individual compounds were detected including some heterocyclic N-compounds such as carbazole Fig 8. It is worth noting that a number of methylated PAH were detected including several other methylchrysenes at levels up to 7 ng/cigarette. The 5-methoge isomer is the most toxic to these isomers.

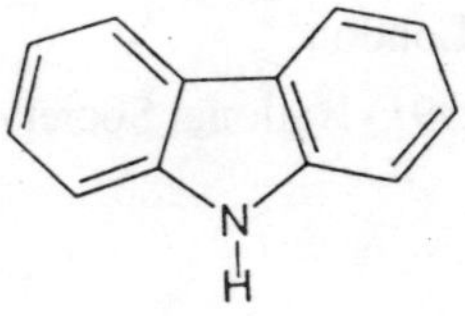

N
H

Carbazole

Fig 8.2 The sliudure of carbazole

Metabolism of aromatic compounds. A principal route in the metabolism of aromatic hydrocarbons is the formation of arene oxides which then isoxerize to phenols (Fig 8.3). In mammals, including man, ingestion of these substances induces changes in the P-450 cytochrome enzymes within the endoplasmic reticulum. This is principally sited in the liver, but also in the kidneys, lungs and intestines. Oxides and phenols formed in this way become conjugated with water-solubilizing molecules and are then excreted.

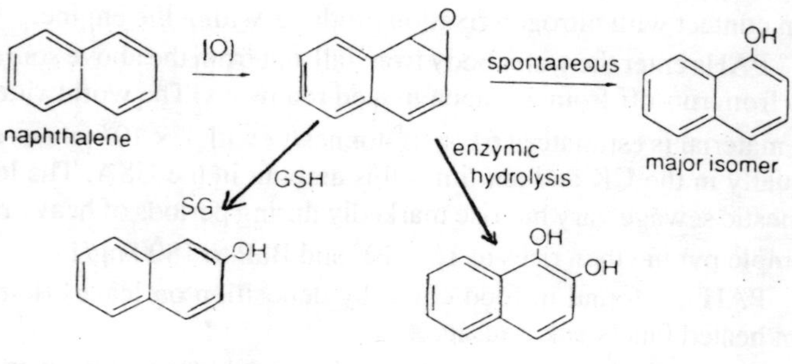

Fig 8.3 Oxidative metabolism of aromatic compounds.

REFERENCES

International Agency for Research on Canur (IARC) (1983). PAH. Part I *Chemical and Environmental and Experimental Data*, 32, WHO, Lyons.

Moriarty, F. (1988) *Ecotoxicology : The study of Polutants in Ecosystems*, Academic Press, London.

Pollution Handbook (1991) National Society for clean Air, London.

CHAPTER 9

Assessment of Environmental Risk

When a toxicant enters the environment many things can happen to it. It may decompose fairly rapidly to form essentially nontoxic products, any harmful effects being localized in time and space. However, the toxicant or toxic derivatives may persist. Then, the extent and degree of environmental damage will depend upon the properties of the chemicals involved and the nature of the affected ecosystem. In this chapter, the properties of the terrestrial and aquatic environments will be outlined, with particular reference to those factors which determine the balance between persistence and degradation of toxicants and those which influence their dispersal through the biosphere. On this basis, the problems of monitoring toxicants and of assessing environmental risk will be discussed.

Toxicants in the Soil

Soil Chemistry and Toxicant Persistence

Soil characteristics affecting toxicant persistence include particle size, mineral and organic matter content, hydrogen ion concentration and microbiological activity. In general, the smaller the soil particles the longer toxic substances will persist. This is because small particles provide a large surface area for absorption of chemicals, a process which usually has a stabilizing effect. On the other hand, adsorption of toxicants reduces their availability to organisms and this may compensate for the increased persistence. For soils of a given particle size, the soil type can be shown to influence persistence, which generally decreases in the following order: organic soil, sandy loam, silty loam, and clay loam.

Probably the two most important inanimate components of soil in relation to toxicant persistence are organic matter and clay. The organic matter consists mainly of humic compounds which have a very high cation exchange capacity. Carboxyl, amino and phenolic groups provide sites for hydrogen bonding with toxicants. Sodium humate has detergent-like properties and can help to solubilize water-insoluble compounds such as DDT.

Clay is the name given to the smallest soil particles (about 0.002 mm in diameter). Clay soils are defined as those with more than 40% of their particles of this size. Clay and organic matter are often associated in soil colloids which can absorb toxicants strongly. The degree of adsorption depends upon the soil pH, moisture content, mineral ion content, temperature and any other factors which influence the physicochemical state of the colloids or toxicants. The electrical potential of the colloid surface is particularly important.

Colloid adsorptive capacity is not the only property influenced by the soil ion content. Hydrogen ion concentration (pH) controls the stability of minerals, soil ion exchange capacity, the rates of chemical reactions and microbial growth and metabolism. Cations affect soil structure by influencing flocculation and dispersal of the colloids. Some metal ions, e.g. those of iron, aluminium and magnesium, can act as catalysts, accelerating breakdown or transformation of susceptible toxicants.

Metabolism of Toxicants by Soil Organisms

Much of the degradation and modification of chemicals in the soil is due to bacteria and fungi. These processes are facilitate by any factors which promote metabolic activity of the micro-organisms. Optimal growth temperature, pH and nutrient availability are particularly important. In some cases, may be required, while in others its absence may be necessary. Before taking any steps to promote microbial transformation of toxicants, it is important to ensure that the products of transformation are not more toxic than the original chemicals, as, for example, in the conversion of inorganic mercury to methyl mercury Soil-dwelling invertebrates, such as mites and earthworms, may contribute to the chemical modification of toxicants. Free enzymes in the soil may also have a role to play. These enzymes come from dying organisms, plant roots and excreta. Since evidence suggests that less than 50% of the known decomposition of pesticides in the soil can be attributed to micro-organisms or nonenzymic chemical reactions, the contribution to this made by invertebrates and free enzymes may be considerable.

Toxicants in Natural Waters

Water Chemistry and Toxicant Persistence

Most of the world is covered by seawater, with a salinity of 35 parts per thousand. Most of the remaining water is classified as fresh water, though rarely as pure water. The solvent power of water ensures that it usually contains dissolved carbon dioxide, oxygen and nitroge, and metals such as sodium, magnesium, calcium and iron.

Water soluble toxicants are rapidly transferred to natural waters, either by leaching from the soil or by precipitation from the atmosphere. Their removal from the water then depends on their other chemical properties. Some may decompose spontaneously or volatilize. Others may form insoluble salts which precipitate, and are incorporated into sediments. Adsorption on to particulates may lead to similar consequences, or it may facilitate ingestion by filter feeders. Uptake by aquatic organisms may be followed by metabolism of toxicants to derivatives of either greater or less toxicity and their subsequent accumulation or excretion. Since many aquatic organisms have the ability to concentrate solutes, without any obvious damage to themselves, they may act as toxicant amplifiers, making the toxicants available to predators at dangerously high concentrations. The predators may accumulate even higher tissue concentrations then the prey because they consume large numbers of them. The process of toxicant amplification can continue to the top of any food chain, provided that the toxicant is chemically stable and has some property, e.g. lipid solubility which inhibits its excretion. Food chain amplification of this kind is observed with, chlorinated hydrocarbons (See Pesticides) and methyl mercury (see Toxic Mekes).

Persistence of toxicants in natural waters reflects their chemical properties. Hydrophobic substances, with appreciable lipid solubility, tend to be the most resistant to spontaneous chemical modification or metabolic change and have the greatest tendency to accumulate in organisms. However, they also tend to accumulate at the water surface. This facilitates their loss by volatilizaton and exposes them to ultraviolet radiation from the sun which may promote their decomposition.

Metabolism of Toxicants by Aquatic Organisms

The role of micro-organisms in the aquatic environment in degrading toxicants is at least as important as in the terrestrial environment. Not only are they involved in the metabolism of toxic substances but some produce toxins of their own. In addition they may cause the death of other organisms as a result of localized oxygen depletion. This accompanies bacterial decomposition of the oxygen demanding wastes which occur in sewage, or effluent from food processing plants, paper mills, tanneries, slaughter houses and intensive animal production units. The effects of oxygen depletion may be aggravated by toxicants, e.g. amines and hydrogen sulphide which are the products of anaerobic breakdown of these wastes. Similar effects of oxygen depletion and toxicant production accompany the decomposition of algal blooms following excessive eutrophication.

Eutrophication

Eutrophication simply means enrichment with nutrients, and its occurrence in natural waters is therefore essential for the maintenance of life. Problems arise when eutrophication is exessive. This state is sometimes referred to as cultural eutrophication. Discharge of nutrients (e.g. nitrate, phosphate) or limiting organic compounds, (e.g. vitamins) in sewage, industrial effluents or runoff from fertilizer-rich agricultural land, can result in such an excess of nutrients in inland lakes and ponds that certain algae flourish abnormally. These algae include the blue-green species *Aphanizomenon flos-aquae, Anabaena spiroides*, and *Oscillatoria rubescens*. Initially, algal photosynthesis oxygenates the water but, as nutrients run out and the algae die, their decomposition is accompanied by oxygen depletion which leads to almost complete elimination of the normal fauna. The production of toxins by some of the blue-green algae serves to accelerate the process. Although generally characteristic of inland waters, it may be that excessive eutrophication is in its early stages in some semi-enclosed marine systemic, such as the Baltic, the North Sea and the Mediterranean.

Salinity

Salinity may be an important toxic factor in the aquatic environment. Many fresh water organisms are killed if their habitat become brackish. The converse is also true, but of less general importance since the tendency of water is always to acquire salts rather than to lose them. The main effect of increasing salinity are due to the associated increase is osmotic pressure of the water. Hence, they are very similar to the effects of dehydration.

Salinity may also be a problem in the terrestrial environment as it affects irrigation of agricultural land. Irrigation water leaches salts from the soil and, if this water is recycled, the salt concentration may become high enough to inhibit crop growth.

Effects to Temperature and Weather Conditions

Temperature

Raising temperature increases the rate of volatilization of chemicals, their water solubility and the rates or reactions in which they are involved. Within limits, it also increases the rate of uptake of toxicants by living organisms and promotes metabolism. These limits vary from one organism to another, but usually approximate to the temperatures for optimum growth. Thus, psychrophiles respond best at temperatures below

20°C, mesophiles between 20 and 45°C and thermophiles between 45 and 65°C. In plants, translocation increases with temperature, as does to the atmosphere of volatile compounds in transpiration or through the cuticle. In animals, increased superficial circulation may facilitate loss of toxicants through the skin, lungs or gills, though reduced kidney excretion may counteract this.

Increased environmental temperatures are usually the result of increased radiation in the form of sunlight. The ultraviolet content of this radiation promotes the breakdown of exposed toxicants, but may also have mutagenic effects by damaging DNA in exposed cells. These mutagenic effects may be beneficial, if they destroy harmful viruses or bacteria, or they may be detrimental, if they cause tumours. Ultraviolet radiation also facilitates vitamin D synthesis in exposed skin. Vitamin D is essential for adequate absorption of calcium from the gut in developing animals and, hence, for proper bone formation.

Another cause of increased environmental temperatures in aqueous habitats is the use of water as a coolant in industrial processes. Used coolant water may have a temperature 12°C higher than the river, stream or bay to which it is returned. One of the most bovious consequences of the discharge of warm water is a decrease in the amount of dissolved oxygen near the discharge point. This is because hot water cannot dissolve as much oxygen as cold water and, being lighter than cold water, forms a blanketing layer on the surface. Increased water temperature also increases the respiration rate of marine organisms. Thus, they require more oxygen in an environment which contains less than before. As a result, they may suffocate. Hot summer days may have the same effect.

Discharge of warm from cooling systems may have more subtle effects by giving false temperature cues to aquatice organisms, hence causing inappropriate behaviour for the time of year, e.g. migration and spawning may be adversely affected. Many molluscs spawn within a few hours after their environment reaches a critical temperature. Normally this will coincide with suitable conditions for larval development. If it occurs at the wrong time, fertilization may be inadequate or larval maturation may fail. Similar considerations apply to most other organisms.

Even organisms which are normally tolerant of elevated temperatures may be adversely affected by warm water discharges.

In this case, the ill effects may be traced to over-rapid changes in temperature which do not allow time for acclimatization. Other ill effects may result from additives, such as chlorine or, in the case of nuclear power plants, from the pesence of radionuclides. Additives are put in to prevent bacterial growth and sometimes to prevent corosion in the pipes.

Minimizing the effects of warm water discharges can be accomplished in one of two ways. Either cooling towers may be used to remove excess heat before discharge, or the warm water may be temporarily stored in a cooling pond. The former process is more expensive to install, but it does present the possibility of using the heat for other purposes instead of releasing it wastefully into the environment.

Weather conditions

Precipitation, in the form of rain, hail or snow, may carry down toxic substances from the atmosphere and deposit them on exposed surfaces. Absorption of such toxicants by plants will be favoured by the wet conditions, but other toxicants may be washed off. Within limits, increase in soil water by precipitation increases inherent biological activity and, therefore, promotes the metabolism of toxic substances. Flooding kills many organisms, but facilitates anaerobic bacteral transformations. It also releases may chemicals from binding to soil colloids, resulting either in their loss by runoff to surface water or their transfer to plants or animals.

Precipitation is associated with increased air humidity, which reduces volatilization of chemicals from the soil and from plants. Guttation in plants, the exudation of water droplets from the leaves, is restricted under these conditions. If the humid atmosphere is in contact with dry soil, water soluble compounds will diffuse with absorbed moisture towards the deeper soil layers. When the soil is moist and the air is dry, this process will be reversed.

Any increase in air movement increases the loss of volatile chemicals from exposed surfaces. If the air movement is sufficient, it may remove surface deposits of particulates or spray droplets. Where leaves are affected, closure of stomata may decrease uptake of toxicants.

Distribution of Toxicants in the Biosphere

No matter how toxic a substance may be, its presence in the environment is of concern only if it affects, or is likely to affect, living organisms. In most cases, this depends upon its uptake by certain of these organisms, and subsequent distribution through the biosphere by way of predator-prey relationships in food webs. A number of attempts have been made to incorporate this knowledge into a predictive model in order to assess the likely risk from new toxicants in the environment. The simplest approach is to consider toxicant distribution as a series of partitions between environment and organism, organism and predator, and so on. Given a constant level of toxicant in the environment, a stable ecosystem and sufficient time, a steady state distribution should be attained. Under these circumstances, each organism in the contaminated environment will contain the toxicant in proportion to a characteristic partition coefficient between itself and the environment, no matter how many intermediaries there may be. Furthermore, the localized toxicant concentration in specific parts of the organism will be nearly constant, though the total will alter with the size of the part. Again partition coefficients may be determined to defined mathematically the distribution between the tissue or cell compartments. Since metabolities partition independently of the parent substance, metabolism of a toxicant may serve only to distribute derivatives throughout the affected organism or, after excretion, throughout the environment. Metabolism cannot reduce levels of toxicant in an organism exposed to a constant external concentration.

The partition approach has been applied successfully to mathematical analysis of closed laboratory systems, but extrapolation to the natural environment has proved difficult for the following reasons. Levels of toxicants do not normally remain constant. Living organisms go through developmental stage of varying susceptility to different chemicals and with varying capacities for their absorption. Short lived organisms may not survive long enough to equilibrate with the environment. Animals may migrate from contaminated environments to clean environments and back again. Plant seeds may be dispersed or lie dormant in the soil. Mutations may occur, and toxicant resistant organisms appear. Because of these difficulties, it seems likely that mathematical models of toxicant distribution in the biosphere will not be of value in environmental planning and management for some time to come. Their chief value at present is probably in generating quantitative hypotheses that can be tested by experiment or observation.

Monitoring Environment Toxicants

Chemical Analysis

Until such time as accurate predictions of the fate of toxicants in the environment can be made, reliance must be placed on well-designed monitoring schemes to provide the necessary information for appropriate steps to be taken in the prevention of environmental damage. Monitoring requires good methods of chemical analysis. Often these are available, but it may not be clear what hemicals the analyst should try to detect. In addition to the suspected toxicant, there will be its metabolites or varying toxicity. There may also be synergists and antagonists. Having decided what to look for, the analyst must know what levels are to be regarded as dangerous and adjust the sensitivity of his methods appropriately. Unnecessary analysis of trace quantities of chemicals can be an expensive waste of manpower and materials. However, establishing safe limits for toxicant exposure is not easy. The natural tendency is to define limits primarily in relation to direct risk to human beings. This may include risk to agriculturally important plants and animals, economically important fish, domestic pets and garden plants. However, their are many other organisms, each playing a part in the total word ecosystem, and our knowledge is as yet insufficient to predict the consequences of partial of complete elimination of many of these species. Thus, while practical necessity makes it essential to define safe limits in relation to those organisms of which most is known, such limits must be open to modification as further knowledge is gained. In cases where no absolutely safe limits can be defined, it may be necessary to establish levels below which risk, though present, can be regarded as acceptably small. This raises the question of what constitutes an acceptable risk. In some cases, e.g. the carcinogens in cigrette smoke, real or imagined benefits associated with the toxicants may make quite a high risk acceptable to *Homo sapiens*.

Even if the chemical analyst knows what to look for and what concentrations are significant, there remain problems of sampling and extraction. Sampling may be random throughout the environment, or selective or a mixture of both. The sampling programme must be designed to the satistically sound where comparisons are being made. Sampling methods must be chosen which are effective in preventing loss of material between collection and analysis, and which avoid contamination of samples with extraneous material. Time between sampling and analysis should be kept to a minimum, or well-established methods of sample and toxiant preservation used to prevent post-sampling changes. Having obtained adequate samples, the analyst must extract the toxicants present. Whatever extraction method is used, some toxicants will be left behind. The question arises as to whether the unextracted toxicants can be safely ignored. The answer to this depends on the relationship between extractability and biologican availability. Normally the two will be closely related, but this may not always be the case. In dealing with higher organisms, it may often be better practice to extract selected parts rather than the whole organism. This is because a high localized concentration in one tissue may be masked when results are expressed on a whole organism basis. How to express results are always a difficult decision. Not only can the amount of toxicant be expressed per organism, or per tissue, but it can be expressed per cell, per gram wet weight, per gram dry weight, per cm^3 body, tissue or cell water, etc. Similar problems arise with environmental material.

Perhaps the most important question for the chemical analyst is what is the minimum quantity of toxicant that can be detected reliably, since this is a measure of the technical limitations that are imposed on environmental monitoring. Awareness of these limitations is essential, because the analyst can never prove that toxicants are absent from the environment but can only demonstrate their presence at levels which respond to his techniques.

Biological Analysis

If assessing hazardous level of chemicals in the environment is difficult, assessing environmental deterioration is even more so. The first problem is in establishing what constitutes normality in the area of interest. This requires what has been called a base-line survey. Ideally, it should involve a control area identical to the environment at risk, but finding such an area may will be impossible. Thus, one has to settle for a similar area, with the error that this entails. Thereafter, a survey is made of the species inhabiting the area and of their relative abundance. It is important that the methods used for assessment are objective and of similar efficiencies in both locations. Transects are often used for studying the vegetation and less mobile animals. They are particularly useful if they can be aligned to a known gradient of toxicant. Otherwise, random quadrats are preferable. Rare species, because of their rarity, may be overlooked in such surveys. This is unsatisfactory because they may be particularly susceptible to toxicants. However, from the information gained on species and their relative abundance, a diversity index may be computed. If the index is low, this is often regarded as a sign of environmental stress but it may reflect other factors, including the homogeneity or heterogeneity of the habitat, its variability with time and rates of predation. Thus, spatial and temporal fluctuations of each species in the control area and the are at risk must be determined to establish their relationship to normal environmental factors and their biological interactions. In this way, a total picture of both areas will be obtained and differences may be observed. The significance of these differences must now be determined and the following questions answered: how many species have to be absent from the contaminated environment relative to the control, and for how long, before damage can be established with certainty; can damage be inferred where species have not disappeared from the contaminate area although their numbers have been significantly changed?

CHAPTER 10

Individual Action to Save the Environment

Introduction

During one hour about 500,000 tonnes of coal comes out of the mines of the world.

Over two million barrels of crude oil (i.e. about 300,000 tonnes) is extracted.

Nearly 200 million cubic metres of natural gas is collected.

One billion kWh of electricity is transmetted.

These estimates from Energy-Food-Environment, (1987 Vadav Sunil) emphasise the fact that it is only we - the ultimate consumers — who can solve the problems of resource depletion ; and of the pollution created by the use of those resources which increasingly threaten our lives. We as individuals can reduce the build-up of chemicals in the environment by taking small precentors and by modifying our lifestyles. The problems which result from the mass production and consumption of products would be eased if we simply reduce the quantities we use.

The Impact of Different Living Standards

The Western industrialised countries have the most to answer for when it comes to the environment because they consume and pollute the most. The poorest people in the third world, living in countries where the gross annual national product may average no more than $250 a head, cannot do much more harm to the environment than their own muscle allows. They build their houses and shelters with their own labour and using materials found locally, which seldom involves any wider harm. They find and carry their fuel from natural and renewable sources; fetch and carry their own water; crop and cultivate their own food; and return their wastes and that of their animals to the soil from which their food is grown. At worst, they can only abuse their immediate environment, and only that if their number, or the number of their livestock, increases beyond levels which that environment will sustain. If that happens, the abuse is brutally self-correcting: they starve.

The western industrialised countries are at the other end of the scale. Their average national product may be between 50 and 100 times that of the world's poorest. Each one of them is responsible for at least 50 times the resource depletion and pollution of third world inhabitants. They take, and with their wealth are able to buy, everything they can lay their hands on. They throw back into the environment any part of it which they cannot use easily or dispose of cheaply. Because of the rate at which draw on resources, the environmental impact of the 56 million people in the United Kingdom equals that of at least 2800 millions living at world subsistence level.

World population now exceeds 5000 million: but if you multiply the number in each nation by the extent to which its living standard exceeds that of the poorest countries, the impact of those 5000 million on the world resources and environment is the equivalent of 53,000 million living at subsistence level.

Of course, the earth lacks both the space and subsistence level resources to support anything like 53,000 million people. It only sustains their equivalent, in the very different living standards of the existing 5000 million, because our skills have allowed us to harness to our advantage resources inaccessible to the poorest. By using those resources we are able to crowd ever larger numbers of people together in ever smaller spaces without risking their decimation by disease; and to multiply and use the earth's natural yields to feed, clothe and water them. But most of those resources are finite and we are exhausting them at a terrifying rate. Energy is the key to all of them; and though it took 400 million years of organic life to accumulate the earth's fossil fuels, we look set to burn them in less than 400 years.

The extent to which we have crawled out on to an increasingly fragile and environmentally dangerous energy limb, is illustrated by a simple comparison. In 1800 farmers in England considered that they were doing quite well, using subsistence methods, to grow and harvest between 3 and 5 hundredweights of barley or wheat on an acre of land (376 kg to 627 kg per hectare). With modern farm machinery, fertilisers, pesticides and weedkillers, farmers worldwide would now expect to cultivate between 3 and 6 tons to the acre (7526 kg to 15,052 kg per hectare). Without these advances, starvation would long since have kept in check national population increase along with world population increase; but starvation will still be there for many millions if energy usage has to be savagely reduced in order to preserve an atmosphere which will not destroy us, or if energy sources are exhausted.

This chapter focuses on specific changes in the way we live at home which can save resources and environmental damage; and on steps we can take to minimise the impact on ourselves of the damage which has already occurred.

It also includes a survey of the background logic for change and for taking avoiding action. Stating the logic for personal action is particularly important in the environmental field. Given the logic, individuals may well perceive, in their unique circumstances, environmentally helpful measures which no general survey could contemplate. Several specific environmentally beneficial actions and activities have been offered by way of examples.

The guiding philosophy is everything we do to save resources and environmental damage is to the good, however trivial individual items may seem; and while our actions may involve large numbers of small contributions towards containing environmental damage, this will enable many of the large ones to take care of themselves.

People in the richest nations, who consume most and pollute most, should take the lead. Several billion people living on earth have so little that they are bound to seek more. But they set the goalposts for what 'more' means.

A review of specific environmentally beneficial measures is given in the following text.

Energy Use

We can reduce the amount of energy required to manufacture products simply by using fewer of them, avoiding their waste and using substitutes which require less energy.

We can significantly reduce the amount of energy which we burn directly. In part, that involves burning less-be it electricity, gas, oil, petrol, coal or any other solid fuel; in part, it involves using the most efficient form of energy and; where possible, that which has the least damaging side-effects on the environment.

Fuel could be burned very efficiently if we only used heat and only burned fuel to produce it where the heat is needed. It is possible to design domestic and industrial fuel-burning appliances which as much as 85 per cent of the heat yielded by the fuel can be delivered for heating. Only the remainder is vented as waste – along with any potentially harmful combustion products.

The energy in fuel mostly has to be converted into mechanical energy before we can use it and the physical laws make that conversion process wasteful. Machines may be driven by gases, vapours or liquids. The vast proportion of the energy we use comes from burning or exploding fuel to produce gases or vapours to rotate steam or jet turbines, or move pistons in steam or oil engines.

Electricity

The overwhelming proportion of the electricity made worldwide is generated from steam-driven turbines. The steam is produced in boilers heated by burning fossil fuel – coal, or gas – or nuclear fuel. Natural energy sources such as wind, water and solar power are unlikely ever to contribute more than a tiny fraction of the total quantity of energy needed. It is an illusion to believe that natural energies can solve our problems.

Megawatts always feature in any discussion of electricity, so it is helpful to define this term. A domestic iron is generally rated at 1000 watts (one kilowatt – k W). The watt is the measure of the quantity of electricity required to power the appliance, so the higher the number of watts the more it consumes. If an appliance rated at 1000 watts is switched on for an hour it will burn one kilowatt hour (h W h) of energy. One kWh is the unit of electricity by which our electricity bills are calculated.

A megawatt (MW) is shorthand for a million k Wh; so a power station which generates 1MW will produce enough electricity to keep 1000 1 kW electric irons going for an hour. When the performance of different types of power station is compared, that of 1000 MW stations is generally used as a standard.

Some countries generate a far higher proportion of their electricity with nuclear power. France leads the field with around 65 per cent nuclear generation; Belgium (60 per cent) and Sweden (50 per cent) come second and third. Countries such as Norway, which have the right natural conditions, generate a large proportion of their electricity from hydro-electic schemes. But with a population of around three million. Norway's total energy needs are low. As far as the rest of the world is concerned, dependence on fossil fuels is fairly typical.

Power stations fuelled by coal, oil, nuclear or other heat energies are inflexible because they must first raise steam before they can generate, and because frequent wide temperature fluctuations spell a very short life for furnaces and boilers. They cannot be started up at a moment's notice, nor can they be regularly started and stopped economically. They are shut down periodically, in carefully controlled conditions, for maintenance and repair, but are essentially designed to keep running all the time.

Generally speaking, a sufficient number of them are kept running to provide enough electricity to meet the average need – the base load.

Stations which can be brought to full power very quickly – diesel, gas turbine, hydro-electric or pumped storage – meet peak demands. In many countries, however, base-load capacity still significantly exceeds that required round the clock. So although great skill and ingenuity is used to balance electricity generation and limit waste, fuel burned in base-load stations may still be entirely wasted as generating continues through periods of very low demand.

The effects of electricity generation on the environment are best illustrated by comparing the quantities of waste which flow annually from different types of 1000MW power stations.

The wastes produced by burning coal for a year in one 1000MW station are in the region of:

1. 7–8 million tonnes of carbon dioxide.
2. 200,000 tonnes of sulphur oxides (dependent on the amount of sulphur in the fuel).
3. 30,000 tonnes of nitrogen oxides.
4. 1000 tonnes of chlorine.
5. 1000 tonnes of heavy metals – zinc, boron, vanadium and arsenic.
7. 250,000 tonnes of ash – fly ash.
8. Significant quantities of radioactive metals and gas (radon) naturally present in coal.

Filtering fly ash out of chimney emissions is relatively cheap and virtually 100 per cent effective. But control equipment for the sulphur and nitrogen oxides, which respectively cause acid rain and contribute to atmospheric heating and high atmospheric ozone damage, is vastly more expensive.

Even control equipment will only recover around 90 per cent of the sulphur and 70 per cent of the nitrogen oxides, and even those oxides which can be contained and removed still have to be put somewhere. A more worrying aspect is that it is not considered possible to prevent the venting into the atmosphere of carbon dioxide and the other substances.

The environmental implications of coal burning are therefore very bleak, and substituting low for high sulphur fuels is hardly an answer to any of them.

Oil-fired power stations have a slightly happier environmental face. The emissions from a 1000MW oil-fired station lie in the region of:

1. 6–7 million tonnes of carbon dioxide.
2. 84,000 tonnes of sulphur oxides.

3. 28,000 tonnes of nitrogen oxides.
4. 6000 tonnes of ash.

Control equipment eliminates the ash; it could reduce the sulphur and nitrogen oxides by the same percentage as for coal. The problems of cost, disposing of the results, and continued carbon oxide emission are identical.

A 1000MW Natural Gas-fired power station has the cleanest fact:

1. 4–5 million tonnes of carbon dioxide.
2. 20 tonnes of sulphur oxides.
3. 25,000 tonnes of nitrogen oxides.
4. 4 tonnes of ash.

As far as nuclear power stations are concerned, obvious considerations of safety make it essential that all radioactive nuclear wastes be contained. Hypothetically at least, nuclear power still has two commanding environmental advantages over all fossil fuels. First, the volume of waste created is physically very small. Quantities are not such that there is no reasonable prospect of containing it, as is the case with burning fossil fuel. Second, fuel does not have to be burned in oxygen in a nuclear power station, so this type of electricity generation does not involve an inevitable and continuing increase in the level of carbon dioxide in the atmosphere. It has been calculated that a nuclear power station saves 99 per cent of the carbon dioxide which would be emitted by a comparable coal-fired station, even allowing for the energy and resources used to build it and produce its fuel.

If our electricity came mostly from nuclear rather than fossil fuels we would not need to worry about the greenhouse effect.

But most of our electricity does come from fossil fuel; and there is no prospect of a change to nuclear power in any time-scale that would spare us the environmental consequences of continuing to burn fossil fuel.

One obstacle to substituting nuclear power is nuclear wastes. The quantity of nuclear waste is sufficiently small to make containing it a realistic possibility; but we know that intensely radioactive wastes (the smallest part in volume) must be safely contained for thousands of years. Nuclear scientists are satisfied that it can be done – by casting reprocessed waste in glass and burying it, glassified, in stable layers below the earth. They can point to a natural illustration of the stability of the process: freak geological circumstances caused a half million year-long chain reaction in uranium deposits at Oklo in the Gabon in Africa, and its high-level wastes have not migrated in a billion years. But people do not believe nuclear scientists.

The fear is understandable. We are all haunted by the Russian catastrophe at Chernobyl, and by lesser accidents elsewhere. Nuclear radiation is invisible and inescapably linked in all our minds with cancer, still one of our greatest fears. No amount of lecturing about the inherently dangerous nature of the Chernobyl reactor, the inherently safe nature of other very different reactors, or the safely of waste disposal, is likely to displace such fears.

The other obstacle is entirely practical. No nation has enough resources to build nuclear-generating capacity sufficient to substitute nuclear for fossil fuel in time; and many developing and the third world countries do not have, and could not afford, an electricity distribution network which would allow the use of nuclear power stations, even if they could build them Widespread use of nuclear power might stave off a fossil fuel-generated environmental catastrophe in our own lifetime, perhaps to the risk of some hazards now and more in the future. But it is too late to think of substituting nuclear for fossil fuels – although it is not too late to think of reducing electricity consumption so that existing nuclear capacity covers more of it.

Details of the waste produced in one year by a 1000MW nuclear power station – note the contrast with that from fossil fuels – complete the picture. The approximate quantities, which vary a little between different types of nuclear reactor, are as follows:

1. 1150 kg of high-level long-life radioactive wastes, requiring long-term containment. After fuel reprocessing and glassification these reduce to a solid 3-metre cube.
2. Between 550 and 1225 cubic metres of medium - to low-level short-life radioactive wastes – gases are usually safely vented to the atmosphere and solids buried in ordinary ground.

Whatever the source of heat from a power station, the amount of waste heat discharged into the environment through cooling towers or in cooling water is about the same. That is an unavoidable component of the high pressure-low pressure differential described earlier. This heat waste can be reduced if the heat can be applied to a useful purpose. Worldwide, a handful of conveniently located housing estates and horticultural and fish farms also use waste power station heat for space or water heating. But piping and retaining heat in pipes is expensive in terms of cash and resources; piping heat over significant distances can cost more energy than it can save.

People and power stations are seldom good neighbours, and in addition people seldom live in places where the vast quantities of cooling water required by power stations are found. Most power stations have to be sited well away from large populations and from commercial ventures which could use waste heat. So waste heat generally an inescapable part of electricity generation.

Turning primary energy into electricity therefore involves losses of heat up the chimney, along with combustion products, and in cooling water; losses of energy from friction as the machinery is driven; losses of electricity required to drive power station pumps and ancillary equipment; and losses from electrical resistance in transmission lines and in the transformers which ultimately reduce transmitted power down to the voltages which we use. The loss from transmission lines turns up in the radio waves which interfere with car radio reception when we drive under them.

As a result of all these inevitable losses, the electricity which finally reaches us reflects at best between 30 and 45 per cent of the energy which went into making it, whatever the source of that energy. The rest, along with combustion products, is waste.

It is for these reasons that several countries – including Britain, until recently – have insisted that fuels such as oil and gas should not be burned in power stations. They are important feed stocks for chemical industries; and a far higher proportion of their energy can be used if they are burned where their energies are needed, as they conveniently can be, to produce heat or drive engines. Unfortunately, burning oil or gas is also the easiest and cheapest way to reduce power station sulphur emissions. As governments snatch at cheap short-term answers to that problem, more oil and gas is likely to be burned to make electricity, not less.

By the time electricity reaches us – the consumers – it is therefore an inherently expensive and wasteful source of heat. Burning it may be less wasteful than burning coal on an open Victorian fire in which all but about 15 per cent of the energy goes up the chimney; but using it as a heat source is far more wasteful than using gas, oil or solid fuel in a modern stove or furnace, which can deliver between 75 and 85 per cent of the fuel's energy for space or water heating or industrial processes. When used in this way, as little as 15 per cent of the energy in the fuel need be wasted – up the chimney but still with its combustion products, of course.

In the home, the moral is quite simple: electricity should only be used when nothing else will reasonably do. We need electricity for electronic equipment such as microwave ovens, radios, televisions, videos, tape-recorders and computers, but electronic equipment only uses very small amounts. We need it for motorised equipment – freezers, vacuum-cleaners, mixers, blenders and refrigerators (though gas refrigerators are also available) – which also require relatively small amounts. We need it for electric lights; and for electric appliances which, while imposing a heavy load while running, compensate for it with their efficiency, convenience and the short time they take to do their job. Thermostatically controlled electric kettles, enclosed cooking and frying pans, and electric irons are examples.

At the domestic level, none of these is a heavy user of electricity. Moreover, modern inventions have made significant inroads into the amount of electricity they need. Fluorescent electric lamps with thin neon or other gas-filled glass elements burn only 17 watts of electricity to produce the same amount of light as the traditional 100-watt tungsten filament bulb. They cost more; but they usually last around 7000 hours against the maximum 1000 hours of most traditional bulbs. Even if such lamps are only fitted in the lighting sockets most regularly used, they will save significant amounts of electricity in any house.

Microwave ovens are another example (though those fitted with a heating element to give food an oven-baked quality use as much energy as comparable ovens heated by gas, electricity or solid fuel). Most rely simply on the microwave process, which uses radio waves in the radar wavebond. The radio waves heat food from the inside by agitating the molecules in it, rather than from the outside by heating the whole even. Microwave ovens are rated at 750 watts or less, and are vastly more economical than the traditional electric oven, rated between 3000 and 5000 watts, or comparable ovens heated by gas, oil or solid fuel.

Electricity is appropriate for these uses; it is not when applied to heating. If gas, oil or solid fuel is used in a modern appliance, the necessary heat can be gained for less than half the quantity of fuel and combustion products involved when electricity is used.

Lighting

Lighting involves the most constantly insidious use of energy. In any house. One light bulb uses very little energy, but very few of us now manage with only one at a time. Ten 60-watt bulbs consume more electricity than a single bar of an electric heater. Many of us use at least that number for their equivalent in higher-powered lamps for long hours in the day. Switching off all unnecessary lights is partial solution; but only a fanatical energy is likely to do that habitually, and anyway some lamps have to stay on. The answer is to fit lamps which produce the same amount of light for less power.

The neon tube principle has now been successfully incorporated into lamps not much different in shape and size from the ordinary light bulb, which can be fitted into standard light bulb fittings. Typically, such lamps draw 17 watts of electricity for the same amount of light as is produced by a conventional 100-watt light bulb. They are more expensive, of course; but the additional cost is repaid by lower electricity consumption and by the fact that such lamps may last 7000 hours or more against the 1000-hour standard of the ordinary light bulb. Even if they are only installed in the light fittings most commonly in use, a householder who fits such lamps can achieve a measurable reduction in electricity used without further thought.

Refrigerators and freezers

Refrigerators and freezers switched on all the time and virtually ignored once installed. Both play an important part in saving other resources. They help to save waste of perishable foods, and make it possible to grow or buy many foods in bulk and reduce the number of energy-consuming journeys which have to be made to suppliers.

A fridge or freezer will not add much to the electricity consumed by an ordinary household, but the total number in use adds significantly to overall electricity consumption. On average, existing cooling equipment uses between 2 and 3k Wh (units) of electricity a year for each litre of space in it; better design can reduce that to figures as low as 0.2–0.4 k Wh per litre. So it is important to enquire about running costs when any new fridge or freezer is purchased.

However well or wastefully refrigerators and freezers are designed, the way they are used and the basic design chosen make important differences to additional wastage.

Virtually all are now thermostatically controlled. Though switched on all the time, they only run and consume energy when the temperature within them rises above the level at which they are set. Saving energy used by cooling equipment is therefore also a matter of reducing to a minimum the number of occasions when that happens.

All cooling equipment warms up and needs more energy to cool down again when it comes into contact with the outside air, so it is important to keep the door of a fridge or freezer closed as far as possible.

Less obvious is the waste which occurs when ice is allowed build up around the cooling elements in a refrigerator, and over those buried in the walls of a freezer. Ice is a very good insulator. A layer of ice between the inside of a fridge or freezer and its colling elements swiftly reduces the amount of heat which those elements can draw out. When the elements are covered by ice and the internal temperature rises, the thermostat cuts in as it usually does; but the equipment then burns up unnecessary energy as a tries to draw enough heat out through the insulating ice for the temperature to drop back below the level at which the thermostat is set. We save energy if we defrost our appliances regularly, as manufacturers advise us.

Domestic refrigerators have a low energy rating and do not add much to the energy costs of individual households. They are not intended to do any more than cool their contents. Freezers are a very different matter.

The most popular freezer is now the upright variety with hung doors which swing on vertical hinges like a house door. They are convenient because it is easy to see what is on the shelves; but they are very wasteful of energy.

Hot air rises and cold air falls; so every time the door of upright freezer is opened, most of its expensively produced cold air falls out and energy has to be burned to replace it. You may need a good memory or marking system to find your stores in a chest freezer, with a door which opens horizontally at the top, but as far as energy conservation is concerned such freezers are much better. Most of the cold air stays locked in the chest even when it is opened.

Cooking Appliances

Many modern appliances use resources, and cook, far more efficiently and conveniently than the traditional alternative. In the kitchen, the hot plate, or gas stove and the conventional oven. But heating anything on an open gas ring is inherently wasteful because so much heat escapes up past the pan on it; and heating anything in a pan without a lid on is even more wasteful, as the heat then escapes out of the pan as well.

Heating a whole oven in order to heat its contents – and drawing on anything between 3000 and 5000 watts of electricity or the gas or other equivalent – is vastly more wasteful than merely heating the food, as with a microwave oven drawing around 750 watts. In addition, microwave çooking preserves more the food's nutritional value and flavour. Faster cooking and the need for far less cooking water account for the difference.

We can do better, and be kinder to the environment, with modern equipment. Thermostatically controlled electric kettles and enclosed cooking and deep frying pans with built-in electric elements deliver only as much heat as is needed and only where it is needed.

By using the higher temperature and penetrating power of steam to do what boiling water does in an ordinary pan, pressure cookers tackle the open cooking ring job more efficiently.

With a little thought, many households could now manage virtually all their cooking with an electric kettle, a couple of cooking pans, perhaps an electric toaster or grill for occasional use, and a microwave oven – and do it far more efficiently in every sense of the word.

Vaccum-cleaners and electric irons

Some appliances are apparently paradoxial; high-powered versions may be more efficient than low-powered alternatives. Vacuum-cleaners and electric ironing equipment come into this category.

The suction power of a vacuum-cleaner is directly related to the power of the motor which drives the fan inside it; so too is the probable life of that motor. A vacuum-cleaner with a 1000-watt motor used in moderation is likely to last longer and to a pick up dust far more quickly than one with a 250-watt motor which is over stretched. The low-powered one may use more energy and resources than the more powerful one, both in day-to-day running and overall.

Conventional electric irons contrast similarly with motorised rotary irons, usually controlled by a pedal. A conventional iron is generally rated at 1000 watts, a rotary one at 1200. But it is much quicker to do the job with a rotary iron, once you realise that shirts and similar garments are still essentially flat if unbuttoned and ironed that way. The shorter running time saves the extra energy needed to drive the rotary iron. The larger area of a rotary iron's heat element, usually significantly greater than that of a flat iron, also saves wear on clothes. The heat is more widely distributed and, the whole operation also saves wear and tear on the operator. Ironing is quicker, and the blade pressure of a typical rotary iron comes from the pedal, not from the bent back of the operator leaning over it.

Automatic Washing-Machines

The automatic washing-machine can probably be singled out as the most environmentally friendly of all household appliances. It washes clothes more efficiently than most people could hope to do by any traditional method, uses little if any more water than those methods would require, and allows thorough, energy-saving, low-temperature washing. The latter is probably beyond the scope of any traditional washing method, except perhaps that of beating clothes clean against a rock in a river. Centrifugal spin-drying, virtually standard in all modern washing-machines, is far and away the most efficient way of extracting water from wet clothes; it saves drying time and resources if clothes have to dried with a heated drier.

Electronic Equipment

The last group of household appliances is the electronic group, which includes radios, televisions, audio equipment, computers, videos-recorders and clocks. Resources are consumed and waste is generated in manufacturing all of these, of course; but the amount of energy needed to run them has fallen to almost negligible proportions since transistors and micro-circuitry came on the scene.

However, this only remains true if they are powered from a mains electricity supply. Batteries are costly in terms of cash, resources and waste. Many, particularly sealed rechargeable batteries, include highly toxic heavy metals such as cadmium which add just a little more to environmental pollution each time they are manufactured or disposed of. Even appliances which do not have a built-in mains lead usually have a plug connection for a transformer lead; and multi-connecting transformers which cost little more than a couple of battery replacements are available, so a transformer soon pays for itself. Wherever possible, therefore, any electronic appliance which is only powered by a battery should be avoided.

Transport

Most of us use cars or other motorised vehicles to bridge at least some of the gaps which separate our domestic lives from the outside world. Vehicles head the list of desirable possessions which damage the environment. Apart from our homes, our cars cost more, and more to run – in cash, resources and pollution – than anything else we individually own or do.

Burning lead-free petrol saves the environment from unwanted lead, but it does not spare the damage caused by winning oil, processing it into fuel, and burning that fuel. If catalytic converters were installed in every car's exhaust they would not eliminate anything like all the carbon and nitrogen oxides emitted by internal combustion engines. And a long time will elapse before there is any technical improvement. In 1990, Britain's Transport Secretary predicted that there would be no significant reduction in vehicle carbon dioxide emissions before the year 2010. Vehicle engines contribute as much as one-fifth of all Britain's carbon dioxide emissions.

In the lower atmosphere, car exhaust emissions add to the constituents of acid rain and to the damage it causes to plant life and buildings. Unburned fuel and oil and petroleum combustion products contribute to the rising incidence of cancer. Carbon and nitrogen oxides fuel and greenhouse effect. The sun's ultraviolet light reacts with vehicle exhaust gases causing complex chemical reactions which result in the formation of toxic groundlevel ozone. In still, sunny weather in heavily polluted urban atmosphere, that ozone accumulates above the tiny levels which even healthy people can tolerance. Ozone also adds to the greenhouse effect.

Car exhaust emissions migrate into the upper atmosphere and there, along with other man-made pollutants, they contribute to the destruction of ozone. We rely on ozone in the upper atmosphere to shield us from excessive ultra-violet radiation. We can only stand infinitesimally small doses of it in the air we breathe. Our vehicle emissions are helping to defeat all our interests.

The only major improvement which individuals can contribute would be to use their vehicles less, particularly in congested urban areas. But some quite simple things would help a little. The faster we go, the more fuel we burn. The World Wide Fund for Nature has calculated that vehicle fuel consumption would be reduced by as much as 2.4 per cent if road users did no more than observe the 70 mph speed limit. And cars fitted with radial rather than the less common cross-ply tyres use between 6 and 8 per cent less fuel.

Unfortunately, our passion for the motor car is so deeply entrenched that none of us seems likely to give up using it willingly, even if we have an entirely adequate alternative. Would the message sink in if we were told that within six months our vehicles would carry us, like lemmings, over a cliff of environmental destruction? Would we not convince ourselves even then than what *we* were doing did not matter ?

It has not come to that – yet. But the example we set is already critical. Worldwide, billions of people hanker after our way of life. If many of them achieve it we shall all suffer in the end if we continue to place the private motor car at thè apex of that way of life.

One can only write hopefully. In a better world we would walk, cycle, or use public transport wherever possible; we would hire vehicles for short-term special needs, possess only the smallest vehicle consistent with our average needs, use it only when nothing else would do, and then as infrequently as possible and over the shortest possible distance. We would buy cars, if at all, for their lasting qualities, and replace them only when beyond feasible repair. We would so arrange our journeys with friends or neighbours that none of us ever travelled without passengers. Passengers can legally contribute to running costs in the United Kingdom.

Many people are already aware of these facts, but reminders are still necessary. Most of us cause more environmental damage with our cars than we do through any other single activity.

Living with Ozone and the Consequence of Ozone Depletion

The ozone layer never stopped all incoming ultra-violet radiation. Had it had done so no one would ever have got a tan from sunbathing, for it is the ultra-violet light which is responsible. Moreover, some ultra-violet light may be beneficial since its action on the skin is a source of Vitamin D.

Nowadays we usually obtain all the Vitamin D we need by eating oily fish, eggs or dairy products; but the chances are that such food sources were often not available to many of our remote ancestors. For them, the ultra-violet skin reaction may sometimes have been vital to survival Vitamin D may provide the answer to why some races evolved with pale skins. People with light skins which admit more ultra-violet light may have been the only survivors in high latitudes where the sun is weaker.

There is, therefore, nothing new in the presence of ultra-violet radiation. Nor is there anything new in its dangers. On the minus side, it has always been capable of causing cellular injury – cataracts, skin damage and skin cancer – and has always caused it. What we have to understand is that the balance of advantages and disadvantages is changing very rapidly. More ultra-violet light is coming through; we either expose ourselves less to sunlight than previously or we increase our chances of injury. If people expose themselves to sunlight for the same periods in the same places as hitherto, increasingly more of them will suffer serious damage.

It is sensible, therefore, to reduce exposure to the sun; to cover up when the sun in at its most intense – particularly around midday and in the summer when it is least obstructed by the atmosphere. People with skins which are sensitive to sunlight should cover up in any event, and avoid any exposure which leaves their skin perceptible hot and red afterwards. And parents should be particularly careful with children, whose skins are always more vulnerable.

People with dark skins (Asians and Africans) are better protected naturally than those with fair skins; people with fair skins which can easily, however, have some degree of protection once they are tanned. But anyone who wishes to enjoy the sun sensibly will aim to take it in far smaller does over far longer periods. Whether it is already possible to feel the change is a moot point. Some people have however reported a prickly feeling in their skin, not previously experienced, when sunbathing in clear conditions in recent years.

What, then, about the ozone in the air we breathe at ground level? Public authorities should warn us if ground-level ozone reaches dangerous levels. If they do, staying indoors will probably be sufficient protection for the majority; but anyone who has a chest or breathing problem ought now to avoid city-centre and other heavily polluted atmosphere on windless sunny days. Ozone builds up in such conditions and may harm vulnerable individuals even if below levels which are generally thought critical.

Surviving the Greenhouse Effect

First, by increasing the proportion of gases which retain incoming solar heat beyond those levels naturally existing in the lower atmosphere, we are raising average world temperature. The extra energy is increasing the power of world weather systems and altering the behaviour of the climate; it is expanding the volume of existing ocean water and rapidly adding to that volume by accelerating the melting of land based ice fields. Low-lying land on which many people live and upon which many more rely for their food is at risk of flooding, and significant sudden increases in sea levels could occur.

Climatic change threatens production from the world's main existing food-growing areas. It may perhaps be possible to grow food elsewhere in the higher latitudes of the northern hemisphere which have been too cold ; We may not be able to move the essential infrastructures of food production swiftly enough to keep pace with them; and natural life may be unable to adapt quickly enough to follow them.

Second, the greenhouse gases for which we are responsible include compounds which break down when exposed to ultra-violet radiation, and release elements which are exceptionally greedy for oxygen. Perhaps this is increasing the proportion of ozone — itself a significant greenhouse gas – in the lower atmosphere, but decreasing the proportion in the upper atmosphere as man-made compounds migrate upwards.

Essentially, the processes involved are as follows: the vast proportion of the oxygen in the lower atmosphere is in the form on which our life depends. When that oxygen migrates upwards and is exposed to intense ultra-violet radiation in the upper atmosphere, the bond between the two atoms is broken and the atoms reform into ozone – a molecule in which three oxygen atoms are combined. The ozone in turn breaks down and the process is repeated. Ultra-violet energy is absorbed by each reaction and the amount of ultra-violet radiation which continues on down to the earth's surface is thus diminished.

When other gaseous compounds migrate into the upper atmosphere, they also are broken down into their elemental components and may then recombine. Compounds which include oxygen, such as the nitrogen oxides, may interfere with the ozone process by collecting more oxygen when they reform than they release when they break down: an example is nitrogen oxide reforming into nitrogen dioxide. But compounds which carry no oxygen with them, and particularly those whose elements are chemically very active, rebond tightly with oxygen and by so doing consume the resources available for the vital ozone process altogether. Chlorine and fluorine are both extremely reactive elements and both are present in CFCs (the chlorofluorcarbons). The CFCs survive unchanged for very long periods in the lower atmosphere and are now acknowledged to be the main cause of ozone depletion as they migrate to high levels. (For more detailed chemistry see chapter 3)

That part of the ultra-violet radiation which passes through the upper atmosphere then causes reactions in the compounds it encounters at ground level – , the nitrogen oxides emitted by motor vehicle exhausts and general fossil fuel combustion. These reactions results in the release of single oxygen atoms, which combine with molecular two-atom oxygen to produce low-level ozone. Ozone is poisonous to us and most life forms in very small quantities, which is why the low level reaction is also a recognised health hazard.

Through these processes, atmospheric pollution attacks several vital life-support systems simultaneously. The ozone shield limits the amount of the incoming solar ultra violet radiation which reaches the surface of the earth and without it all life on earth would be destroyed. Even thinning of the shield has serious consequences. Ultra-violet light damages animal and plant cells and, in our own case, leads to cataracts and skin cancers; the greater the level of ultra-violet light reaching the earth's surface, the greater is the incidence of this damage. Ultra-violet light also reacts at surface level with pollutants producing higher ozone levels there. Ozone is poisonous to animal and plant life in very small quantities, apart from being a greenhouse gas.

The increasing risk is measured by research. Apart from the ozone holes over Antarctica, investigators have reported a 3 per cent thinning in the ozone layer over the United States and Europe between 1969 and 1986; and measurements of incoming ultra-violet radiation taken at ground level.

Greenhouse Gases

The man-made gaseous compounds which are damaging the environment as we pump them into the atmosphere are present only in minute proportions – parts per million (ppm). However that does not prevent them having profound effects. CFC levels are rising at 6 per cent a year, while those of CO_2 are only increasing at 0.4 per cent a year; and CFCs retain 10,000 times the heat of CO_2 in the lower atmosphere, as well as being the most vicious of the ozone observers above it.

The greenhouse effect of the five gases or compounds which are mainly responsible is therefore best illustrated by multiplying the proportion in which they are actually present by the extent to which their effect exceeds that of CO_2.

Gas	*Actual concentration (ppm)*	*Effect elative to CO2*	*Annual in ncrease %*
CO_2	350.0	× 1	0.4
Methane	1.7	× 30	1.1
Nitrous oxides	0.3	× 150	0.3
Low-level ozone	0.001	× 2000	0.25
CFCs	0.0006	× 10,000	6.0

The warming of the earth's atmosphere threatens to produce more extreme weather conditions more frequently: heat and drought, as well as storm, tempest and flood.

It will pay us, and the environment, to collect and retain rain water for uses for which pure water is not necessary ; to remedy structural defects which make our houses more vulnerable to being ripped open in violent weather; to store articles particularly sensitive to water damage sensibly if there is a flood risk; to top or remove trees which would cause damage or injury if felled by a storm; to clear potentially inflammable undergrowth round our houses; and to review how we might keep the basic essentials of life going if public supplies, services or communications fail.

We have already experienced extreme weather conditions which have left people without food, water, heat, light, shelter, communications, transport, first aid or the benefit of emergency services. The information emerging about the greenhouse effect suggests that such experiences will not be unique in our lives.

Household and Industrial Wastes

Gaseous Waste

Gases created by the fuels we burn in our houses and industries and the way in which we burn them or contribute to their burning have been discussed in chapter 3. Their effect in the atmosphere is also considered. We can only contribute to their reduction in the ways already discussed: the best thing we can do, therefore, is to burn as little as possible, and consume fewer products which demand the burning of energy.

These days, most domestic rubbish is collected in plastic bags, carted off and buried in open landfill tips. If food and other organic waste which will slowly rot is included in that solid waste, dogs, cats or rats will often sniff it out before it is collected.

Any food and organic waste which then remains (and large proportions do) will decompose; but because rubbish is buried away from oxygen, the processes of decomposition will release methane along with some carbon dioxide – typically 70 parts of methane for every 30 parts of carbon dioxide. The increase in the general level of methane released is a matter of growing concern because of its contribution to the greenhouse effect. Methane retains 30 times more heat than carbon dioxide, so it is important to reduce the amount of organic waste which is tipped.

Part of that reduction can come from using up everything which can be used. But if after that you can compost or burn organic rubbish at home rather than send it to the tip, the environment will benefit. Loose composting or burning will produce carbon dioxide; burying rubbish in landfill sites will produce the far more heat-retentive methane, as well as attracting scavengers and feeding vermin.

This is one case in which the pollution you create and can see for yourself at home may be less harmful than that to which you contribute and cannot see somewhere else.

Your neighbors might not see it that way, of course, particularly if they are saving energy by hanging their washing out to dry. So it pays to be considerate.

Municipal authorities do not burn waste because confining it to landfill sites is a great deal cheaper. Incineration has been explored. Resources could be saved with profit if the energy which waste contains could be used.

Practical problems prevent the harnessing of that energy. Waste sources are widely scattered and long-distance transport would dissipate their value; raw waste cannot be stored, and the supplied available at individual locations would suffice only to fuel small power stations. They cannot generate electricity economically because of the energy lost in the process and the high cost of generating equipment. Raw waste would produce heating steam at a profit, but to use that steam we would need incinerators in commercial areas where every aspect of waste burning is unwelcome.

A handful of large local authorities has invested in complex and costly plant which sorts waste and pelletises the combustible part into fuel. Such fuel has 60 per cent of the heat value of power stations if the majority of waste disposal authorities could afford the necessary plant – but the majority cannot afford it.

Solid waste

If you have convenient local disposal facilities into which you can separate wastes such as aluminium cans, glass bottles and paper for recycling, obviously you should aim to use them. Even that requires caution, however: you can easily burn more fuel and resources by driving with supposedly recyclable wastes to a central collection point than will afterwards be saved by recycling them.

Bottle-banks and other general recycling facilities are not viable unless it is economic (in cash and resources) to collect and remove waste sorted and deposited in them. But few of us are equal to the task of calculating whether our end of the exercise makes environmental sense. By and large, if dumping waste in a central collection point involves your making a journey which you would not otherwise make, it won't. All your intended economy will vanish in the resource cost of the journey.

Sorting solid waste for external recycling is only part of the picture. Recycling begins and is at its most efficient at home.

A host of things we buy these days come ready packed in plastic boxes and other containers. All represent manufacturing resources already used by petrochemical and other industries. Cellophane apart, they all take a very long time to decay naturally, and burning them releases several harmful gases. Plastic containers are ideal for storing dry foods in the kitchen and chilled and frozen foods in the refrigerator and freezer: they should be kept for that purpose. It makes no sense at all to throw them away and then buy and use cling-film and plastic containers for the same job – which requires additional resources and involves further environmental harm.

Cartons in which cream, yoghurt and similar substances are sold are useful for plants. Punch a small drainage hole in the bottom and you have an ideal substitute for the flower-pot or seedling container. Many gardeners already use them instead of the plastic or ceramic alternatives, which again swallow up additional resources.

ICI has recently announced the first commercial use of a plastic, Biopol, developed from fermented organic sugars rather than the oil by-products which make up most other plastics. Biopol decomposes completely in soil bacteria when tipped and buried. Decomposition releases carbon dioxide; but this is balanced by the carbon dioxide taken up naturally to produce the sugars. So once Biopal and plastics like it are in general use, dumping them will not contribute to the build-up of plastic litter or carbon dioxide.

Packaging is so basic to the marketing of manufactured goods that whatever happens in the future all of us are still likely to be left with some glass, metal, plastic, paper and cardboard rubbish or containers which we can only burn or throw out. But with a little thought we can all reduce their quantity and by doing that, the additional resources used to provide alternatives. Everything helps, from buying loose rather than prepacked goods to making sure that we re-use the ubiquitous plastic carrier bag rather than accept yet another.

Food Wastes

Food wastes are residues of something which has already cost an arm and a leg in resource terms when we buy it. In the days when some beef cattle were reared exclusively on barley, for example, seven tonnes of barley feed went into producing every tonne of beef. All the animal protein we consume and waste – fish, meat, eggs, milk and their products – involves a conversion of foods and resources. We should therefore recover for our own consumption all we possibly can.

Nowadays, something as simple as the declining use of home made cooking stock contributes significantly to food waste. After we have consumed all the meat we want we can further reduce the waste inherent in what is left by boiling bones or poultry carcases in a pan or pressure cooker and preserving or freezing the resulting stock – in some of those plastic boxes. Some vegetable wastes can also be dealt with in this way.

When we have recovered all we can, our pets may be able to recover more from what is left, saving at least part of the formidable resources swallowed up by the pet-food industry; and though there may then be no choice but to burn or put out for the domestic waste collection any animal residues which remain, virtually all vegetable residues can be composted for use in the garden. Even tea leaves are useful there, particularly for improving soils in which herbs are grown, including soils in window-boxes.

As individuals , we can probably recycle more of our own waste than public services can. There may be a limit to what anyone living in a flat in an apartment black can do ; but everyone can do something, and many of us could do a great deal more than we do. If we reduced the volume of solid waste which leaves our homes, and reduced the proportion of it which can cause further harm, we would be contributing to the advantage of the environment. The ideal rubbish bag is one which contains at most inert substances – ashes, cans, bottles and containers which may rust but will not rot.

Liquid Wastes

Similar principles apply to the liquid wastes we generate and then flush down the drain. The quantity of liquid waste we release relates very closely to the amount of fresh water we use.

On an average each of us uses about 100 litres of water a day. We only use between 10 and 13 litres as drinking and cooking water.

The quantity matters both because it costs energy and resources to pipe clean water into our homes in the first place; and because it costs still more to treat it after we have fouled it and sent it on its way.

But the quality of the waste water we release is particularly important. If we put environmentally harmful substances into water when we use it, it may be impossible to remove them during subsequent treatment, and is in any event costly in resources. We increase the volume of harmful residues which flows into the general environment every time we add harmful substances to our water.

Quantities

As far as the quantity of water we use is concerned, nothing concentrates the mind more than having to pay for your water through a water meter.

Obviously, dripping taps and leaking pipes are to be avoided. Many of us have ingrained habits which are wasteful, such as washing-up under running taps, or washing anything up or away in small quantities when a short delay will allow the same quantity of water to be used for a larger job. Fastidious people may not enjoy the thought of not flushing the toilet every time it is used, or of allowing washing or washing-up to pile up before tackling them. But most of the filtered water we consume is used for these purposes.

We can also save toilet water systematically. The cisterns on most toilet hold a good deal more water than is usually needed to flush them ; cisterns empty and are refilled every time they are flushed. During Britain's 1976 drought the Duke of Edinburgh suggested that everyone should lift the lid off the toilet cistern and put a brick in it, saving water equal to the volume of the brick with each flush. We can achieve the same result, with less risk to the cistern and its equipment, by filling a suitable plastic bottle – recycled of course – with water and sinking it in the cistern. If we do this our water saving is regular, permanent, effortless and considerable.

It also helps significantly if we can avoid the use of quality drinking water for purposes which do not require it. That too is not difficult. The rain falls on all of us, and can provide many of us with a freely available alternative. Plastic roof downspouts with a cut which is conveniently situated can be used to store rain water by placing a large rainwater butt or tank under it, with an overflow back into the surface-water drain. This gives a sizeable store for rainwater off the roof of our house which is topped up every time it rains without any further effort on.

Rainwater is useful in any garden or greenhouse; if the tank levels are right, we can run it through a hose. Rainwater does not contain substances such as chalk or lime which may be present naturally or will have been introduced into your main supply if it is naturally acidic. All public supplies are kept slightly alkaline to prevent corrosion in metal pipework, but as a result acid-loving plants like azaleas never flourish on tap-water.

Public water supplies are also slightly chlorinated to kill off any bacteria which may penetrate them. Chlorine is used for the same purpose, but in larger quantities, in swimming pools.

Chlorine is not beneficial to any living thing; and although the proportions in which it is used in drinking water probably do us no harm – or at least less harm than bacteria would – it may harm plants which are watered over and over again in a greenhouse, plant pot or other restricted environment.

Rainwater has other uses and can be employed for everything from washing down surfaces to cleaning cars. Every drop saved and used spares resources required to provide the purified public supply.

Quality

The quality of the liquid wastes we discharge depends on the substances we add to water before it goes down the drain. For example, almost everyone uses detergents. Many liquid detergents still includes phosphates (used as a water-softener) which cannot readily be extracted from sewage. Phosphates fertilise and encourage the spread of algae and other plants which clog watercourses and reduce water oxygen levels, killing fish and other water animals. They may eventually turn watercourses into foul and lifeless sewers.

Many detergents also include foaming agents to produce a lather. Lather produced in soapy water indicates that there is enough soap in the water to clean, but detergents do not work in the same way and do not need to show a lather at all. Foaming agents persist in water when it goes down the drain and cause foam to build up in watercourses.

Environmentally friendly detergents which do not include phosphates are now available commercially and we should buy them – as well as detergents free of foaming agents, if available. But ordinary washing soda used (in combination with soap for washing purposes) is a very good watersoftner and general cleaning aid. If anything, it does good rather than harm after it leaves us in our waste water.

Bleaching powder (sodium hypochlorite) is potentially damaging. It is a killer – indeed, its germ-killing quality is widely advertised – and again it is the chlorine in the bleach which does the killing. A good deal of that chlorine remains active and goes on killing after we have sent it down the drain.

If the foul water flows into a septic tank it is harmful. A septic tank only functions efficiently if anaerobic bacteria present in all sewage flourish and digest solid waste. If bleach goes down the drain regularly, it kills those bacteria. When that happens, the tank seizes up and requires regular pumping out.

Comparable results can follow the general large-scale emptying of bleach into public sewers. Much public sewage treatment also depends, or used to depend, on natural anaerobic fermentation. If the quantity of bacterial toxins in sewage makes that impossible, more expensive treatment is necessary.

Whatever other consequences there may be, all use of bleach adds to the quantities of chlorine loose in the environment. We needs to remember that chlorine is potentially harmful to virtually every life form and in virtually every form in which it exists. Its commonest compound, common salt or sodium chloride, is least harmful; but no animal can take great quantities of salt, and very small quantities may wither most plants. Another sodium-chlorine compound – sodium chlorate – has long been used specifically as a weedkiller.

More modern manufactured chlorine compounds have become notorious. They include the persistent organo-chlorine pesticides like DDT (dichlor-dipheny-trichloroethane); the PCBs (polychlorinated-biphenyls); and the compounds formed in the chemical reactions which lead to the forming of dioxide. Their devastating long-term effects are best known from the Seveso disaster in Italy, and the American use of the defoliant Agent Orange in Vietnam.

The chlorine in CFCs (chlorofluorocarbons), used as a coolant gas in refrigerating equipment, a propellant (now largely outlawed) in aerosols, and a foaming gas in plastic manufacture contributes both to the greenhouse effect and the destruction of the ozone layer. In its elemental form, chlorine was used as a poison gas in the First World War.

So we face a dilemma: bleach may be very useful at home, but we do no one any good by using it when it can possibly be avoided. Apart from using tiny quantities to wipe off mildew stains there are not many domestic uses of bleach which can be commended. Most of its conventional uses result in its being flushed down the drain sooner or later.

What could we use instead? Germ-killing properties often figure in bleach sales promotion and it may be comforting to imagine loss and drains without bacteria. But we should not be too impressed by that propaganda. Few of us have close enough contact bacteria. We are far more likely to encounter germs if we bathe in water polluted by sewage outfalls, among the bacteria which have survived all the bleach – as we have done for years.

On the other hand, bleach does clean. We have become attached to shiny, clean-looking toilet bowls, and detached from the lavatory brushes which were once used to keep them that way.

There are alternatives to bleach. For other purposes, washing soda may be used as a substitute. Soda is a simple, cheap, long-established alternative which can be used as an aid in a host of washing and cleaning operations. Mildly alkaline, it helps break down grease, and does no environmental harm after it has been used. Nor is soda harmful to us: bath salts are based on it. However, it is advisable to rinse it off the skin if used in concentrated solution.

If substances other than bleach may cause harm when we use them, the safest general rule is to assume that they will also cause harm if put down the drain. That applies to all forms of oil, all chemical solvents, and virtually all waste medicines. Cooking fat and grease are forms of oil which solidify on contact with cold surfaces and may block drains: pour them on your compost heap, burn them, or put them into an empty tin in your rubbish bag.

Solids which will not dissolve should also be burned, or wrapped and put out for refuse collection. They should not go down the drain. Nappy-liners, condoms, tampons, sanitary towels and similar detritus may block the own drains or cause blockage elsewhere. They hinder sewage treatment. If you visit a sewage works you will see for yourself the scale of the equipment and resources devoted to moving and shredding insolubles which people have thoughtlessly flushed away. If you walk along a sea-shore near an untreated set sewage outfall, the environmental impact of such items will be inescapable.

Only soluble organic wastes and cleaning substances declared safe on their packaging should go down the drain. If we all accepted that as an objective, subsequent problems in waste treatment and disposal would be significantly reduced. In particular, more sewage sludge might then be recycled back for use as fertiliser instead of being dumped on land or, worse, at sea.

General Waste

In many respects we have become instinctively wasteful. We buy because merchandise is new or fashionable, even though we already have less fashionable equivalents which are still entirely serviceable. We buy gifts for our children, even when the chances are that those gifts will be put to one side almost as soon as they are opened; and that our children will continue to find pleasure in cherished existing possessions. We buy on impulses sparked by heavy advertising when a little ore thought and research might lead us to an alternative which costs less, and is more serviceable. Product research before purchase is far less widespread than it should be.

We all have habits which inspire waste and we are vulnerable to the advertising subtly and deliberately designed to exploit them. The environment suffers unnecessarily each time they are exploited. If we confined our purchases to articles which we really need, or at least to those value to us is commensurate with the cost and the resources used to produce them, that alone would contribute significantly to our welfare and that of the environment.

The pressure to accept wasteful giveaways is even more insidious than that in the advertising of the goods we buy. How many plastic carrier and other bags and packages do we accept without thought when we go shopping? How many of them might we re-use, over and over again, if we preserved them and remembered to take them with us next time round? How much junk mail do we accept without protest and without any effort to prevent its reaching us?

All this unsolicited and unwanted rubbish costs resources; all of it represents waste; all of it has cost someone something. It is up to us to be less receptive.

Finally, there is pre-packaging. The philosophy of the throwaway package has eaten deeply into our way of life. Almost without thought, we reach for pre-packed items even when the same commodity is available unpacked. The unpacked item frequently costs less and, if perishable, is fresher.

The same health and quality regulations apply to perishable foods whether sold wrapped or loose. Food sold loose has to be fresh to comply with those regulations; food sold packed may comply only because sterilisation, preservatives and other additives arrested decay. It is easy to understand why people who sell perishable foods prefer the easier handling and longer shelf life of pre-packed items. Whether our preference should always be the same merits thought. The environment is suffering from excessive pre-packing.

Appendix—A

Exposure limits of Hazardous Chemicals

The National Institute for Occupational Safety and Health (NIOSH) USA develops, recommendations for limits of exposure to potentially hazardous. It also recommends preventive measures designed to reduce or eliminate adverse health associated with these hazards. These recommendations are then published and transmitted to the Department of Labour, Occupational Safety and Health Administration (OSHA). NIOSH recommendations are published in a variety of documents. Criteria Documents specify a NIOSH Recommended Exposure Limit (REL) and appropriate preventive measures designed to reduce or eliminate adverse health effects.

The following table provides the NIOSH REL or other recommendation for each potential hazard. The current OSHA Permissible Exposure Limit (PEL) or standard is also included.

Definitions of abbreviations and terms used in this publication :

Ca	potential human carcinogen
dBA	decibel, weight according to the A scale, which approximates the response of the human ear
ECG	electrocardiogram
J/cm^2	joules per square centimeter
μ m	micrometer
$\mu g/m^3$	micrograms per cubic meter
mg/m^3	milligrams per cubic meter
mppcf	millions of particles per cubic foot

mW/cm^2	milliwatts per square centimeter
nm	nanometer
NIOSH	National Institute for Occupational Safety and Health
PCB's	ploychlorinated biphenyls
PCDD's	polychlorinated dibenzo-*p*-dioxins
PCDF's	polychlorinated dibenzofurans
ppb	parts per billion
ppm	parts per million
REL	Recommended Exposure Limit (NIOSH)
TCDD	2, 3, 7, 8-tetrachlorodibenzo-*p*-dioxin
TWA	time-weighted average

SUMMARY OF OSHA REGULATIONS AND NIOSH RECOMMENDATIONS

Chemical	NIOSH		
	REL's Recommendations	Health Effect	Remarks
1	2	3	4
Acetylene	{*No exposure* > 2,500 ppm (2,662 mg/m^3	Asphyxia	
Acrylamide	0.3 mg/m^3 TWA	Skin, eye, nervous system effects	Prevent skin and eye contact
Acrylonitrile	Ca 1 ppm. 8-hr TWA : 10 ppm ceiling (15 min) (Skin)	Brain tumors; lung and bowel cancer	Chest x-ray require; first-aid and medical kits to be available during use; prevent skin contact
Aldrin/dieldrin	Ca Lowest reliably detectable level	Potential cancer in humans; produced tumors of the lungs. liver, thyroid, and adrenal glands in animals	Aldrin/dieldrin no longer produced in U.S.; prevent skin contact
Alkanes	All are TWA values : Pentane : 120 ppm (350 mg/m^3); hexane : 100 ppm (350 mg/m^3); heptane : 85 ppm (350 mg/m^3); octane : 75 ppm (350 mg/m^3); Mixtures not to exceed (350 mg/m^3); TWA; All are ceiling value (15 min) singly or mixture; pentane : 610 ppm (1,800 mg/m^3); hexane : 510 ppm (1,800 mg/m^3); heptane : 440 ppm (1,800 mg/m^3); octane : 385 ppm (1,800 mg/m^3); Action level set at 200 mg/m^3 ; for these substances	Skin and nervous system effects singly or mixtures;	None

Chemical	NIOSH REL's Recommendations	Health Effect	Remarks
1	2	3	4
Allyl chloride	1 ppm (3.1 mg/m^3);~TWA ; 3 ppm (9.3 mg/m^3) ceiling (15 min)	Liver,kidney,and lung effects	Urine,blood,and pulmonary function testing required
4-Aminodiphenyl	Ca	Bladder cancer	None
Ammonia	50 ppm (34.8 mg/m^3) ceiling (5 min)	Respiratory irritation	Prevent eye contact
Animal rendering process		Mechanical injury ; burns : heat stress ; infections from biologic agents ; chemical hazards	Guidelines for engineering controls and work practices to reduce injury and illness presented
Antimony	0.5 mg Sb/m^3 TWA	Irritation; cardiovascular and lung effects	Chest x-ray,pulmonary function testing,and electrocardiogram required
Arsenic,inorganic	Ca 2 μ g As/m^3 ⌈ (15 min)	Lung and lymphatic cancer; dermatitis	Chest x-ray required
Arsine	Ca 2 μ g As/m^3 (0.002 mg As/m^3) ⌈ (15 min)	Sudden extensive hemolysis	Workers to be warned of working with arsenic compounds in presence of freshly formed hydrogen
Asbestos	Ca 100,000 fibers/m^3 , over 5 μm in length,8-hr TWA in a 400-liter air sample	Lung cancer ; mesothelioma ; asbestosis	Chest x ray and pulmonary function testing required
Asphalt fumes	5 mg/m^3 ceiling,measured as total particulate (15 min)	Eye and respiratory irritation	Medical surveillance required ; prevent skin contact
Benzene	Ca 0.1 ppm (0.32 mg/m^3) 8-hr TWA ; 1 ppm (3.2 mg/m^3)	Cancer (leukemia)	Prevent skin contact

Chemical	REL's Recommendations	NIOSH Health Effect	Remarks
1	2	3	4
Benzidine	Ca Use 29 CFR 1910. 1010	Bladder, liver, and kidney cancer	None
Benzidine-based	Ca Reduce exposure to lowest feasible level ; replace with less toxic materials	Bladder cancer	Stringent workplace controls and medical surveillance required. Urine monitoring for benzidine suggested
Benzoyl	5 mg/m^3 TWA	Respiratory and eye irritation ; skin effects	None
Benzyl chloride	5 mg/m^3 ⌈ (15 min)	Irritation ; skin and eye effects	Chest x ray and pulmonary function testing required
Beryllium	Ca Not to exceed 0.5 μ g Be/m^3	Lung cancer	Pulmonary function testing and chest x ray required
Boron trifluoride	No exposure limit recommended due to the absence of a reliable monitoring method	Respiratory effects	Appropriate engineering and work-practice controls to reduce exposure to lowest feasible level ; pulmonary function testing required
1.3-Butadiene	Ca Reduce exposure to lowest feasible level	Hematopoietic cancer ; teratogenicity ; reproductive system effects	Appropriate engineering and work-practice controls ; restrict access to areas where 1.3-butadiene is used
Cadmium	Ca Reduce exposure to lowest feasible level	Lung cancer, prostatic cancer, renal system effects	None
Carbaryl	5 mg/m^3 TWA	Central nervous system and reproductive system effects	Workers to be warned of possible effects on reproductive system and to have only minimum exposure during pregnancy ; prevent skin and eye contact

Chemical	NIOSH REL's Recommendations	Health Effect	Remarks
1	2	3	4
Carbon black	3.5 mg/m^3 TWA ; Ca 0.1 mg/m^3 TWA in presence of polycyclic aromatic hydrocarbons	Lung cardiovascular and skin effects ; cancer of the lymphatic bone marrow complex when exposed to carbon black in the presence of polycyclic aromatic hydrocarbons	Chest x ray,pulmonary function testing and ECG required
Carbon dioxide	10,000 ppm (18,000 mg/m^3) TWA ; 30,000 ppm (54,000 mg/m^3) ceiling (10 min)	Respiratory effects	None
Carbon disulfide	1 ppm (3 mg/m^3 TWA ; 10 ppm (30 mg/m^3) ceiling (15 min)	Cardiovascular,central nervous system,and reproductive system effects	Workers to be advised of potential effects on reproductive system
Carbon monoxide	35 ppm (40 mg/m^3), 8-hr TWA ; 200 ppm (229 mg/m^3) ceiling (no defined time)	Cardiovasculr effects	None
Carbon tetrachloride	Ca 2 ppm (12.6 mg/m^3) ceiling in a 45-liter sample (60 min)	Liver cancer	REL based on lower limit of detection at time of document publication
Chlorine	0.5 ppm (1.45 mg/m^3) ceiling (15 min)	Eye and respiratory irritation	Chest x ray required
Chloroethane	To be handled in the workplace with caution	Central nervous system effects ; possible liver and/or kidney effects	Exposures should be minimized due to the structural similiarity to the carcinogenic chloroethanes
Chloroform	Ca 2 ppm (9.78 mg/m^3) ceiling in a 45-liter sample (60 min)	Liver or kidney tumors and central nervous systems effects	None
bis-Chloromenthyl ether	Ca Use 29 CFR 1910. 1008	Lung cancer	None

Chemical	NIOSH REL's Recommendations	Health Effect	Remarks
1	2	3	4
Chloroprene	Ca 1 ppm (3.6 mg/m^3) ceiling (15 min)	Lung and skin cancer reproductive effects	Chest x ray and pulmonary function required ; pregnant workers to be counseled about continuing work with chloroprene
Chromic acid	25 µg/m^3 (0.025 mg/m^3) TWA; 50 µg/m^3 (0.05 mg/m^3) ⌈ (15 min) as noncarcinogenic Cr (VI)	Nasal ulceration	None
Chromium (VI)	Ca Carcinogenic Cr (VI): 1 µ g/m^3 TWA; other Cr (VI): 25 µg/m^3 TWA; 50 µg/m^3 ⌈ (15 min)	Lung cancer, skin ulcers, and lung irritation	Employer must demonstrate absence of carcinogenic Cr (VI) ; x ray required
Chrysene	Ca To be controlled as an occupational carcinogen	Potential for cancer in humans; produced liver and skin tumors in animals	Document also contains control recommendations for ploycyclic aromatic hydrocarbons
Coal gasification plants	—	Various effects depending on substances present ; potential for skin cancer	Extensive work-practice and control procedure recommended
Coal liquefaction		Various effect depending on substances present ; potential for skin cancer	Extensive work-practice and control procedures recommended
Coal tar products	Ca 0.1 mg/m^3~TWA (cyclohexane-extractable fraction)	Lung and skin cancer	Includes coal tar pitch; pulmonary function testing and chest x noys required

Chemical	NIOSH REL's Recommendations	Health Effect	Remarks
1	2	3	4
Cobalt	NIOSH has concluded that there is insufficient evidence to warrant recommending a new exposure limit	Dermatitis ; potential for pulmonary fibrosis	Includes recommendations for engineering controls, work practices, protective equipment, worker education, monitoring, and medical surveillance.
Coke oven emissions	Ca 0.5-0.7~mg/m^3(500-700 µg/m^3) (total particulates) as screening level	Lung cancer	Chest x ray required ; work practices to minimize exposure to emissions
Cresol	2.3 ppm (10 mg/m^3) TWA	Skin, liver, kidney, and pancreas effects	Applies to mixtures to cresols and cresylic acid : prevent skin and eye contact : possible delayed effects
DDT	Ca Lowest reliably detectable level; 0.5 mg/m^3 TWA by NIOSH-validated method	Potential for cancer in humans ; produced tumors of the liver,,lungs,,and lymphatic system in animals	Prevent skin contact
2,4-Diaminoanisole and its salts	Ca Reduce exposure to lowest feasible level	Potential for cancer in humans; produced tumors of the thyroid,skin,and lymphatic system in animals	Prevent skin contact ; engineering and work-practice controls are recommended
o-Dianisidine based dyes	Ca Should behandled in the workplace with caution ;	Potential for cancer in humans ; produced tumors of the bladder.	Substitute less toxic dyes wherever possible
Dibromochloropro pane	10 ppb (0.1 mg/m^3) TWA (NIOSH recommendation 0superseded by OSHA standard promulgated in 1978)	Sterility; renal and liver effects	Regulated by OSHA as a carcinogen

Chemical	NIOSH		
	REL's Recommendations	Health Effect	Remarks
1	2	3	4
3.3 Dichloro-benzidine	Ca Use 29 CFR 1910.1007	Potential for cancer in humans; produced tumors of the liver bladder, and lungs in animals	none
1.1 Dichloroethane	To be handled in the workplace with caution	Central nervous system effects; possible liver and/or kidney damage	Exposures should be minimized due to the structural similiarity to the carcinogenic chloroathanes
Di-2-Ethythexyl Phthalate	Ca Reduce exposure to lowest feasible level	Potential for cancer in humans ; produced liver tumors in animals	DEHP, widely used in the guantitative fit testing of respirators should be replaced with less toxie material such as refined co.
Diisocyanates	All values given in μ g/m^3 and all ceiling values for 10 min (each equivalent to 5 ppb) TWA and 20 ppb ceiling): TDI: 35 TWA, 140 ceiling ; MDI: 50 TWA, 200 ceiling; hexamethylene diisocyanate (HDI) : 35 TWA, 140 ceiling ; naphthalene diisocyanate (NDI) : 40 TWA,170 ceiling ; isophorone diisocyanate (IPDI) : 45 TWA,180 ceiling ; dicyclo-hexylmethane-4,4'-diisocyanate (hydrogenated MDI): 55 TWA 210 ceiling; other diisocyanates to be controlled to 20 ppb ceiling and 5 ppb TWA	Respiratory effects and sensitization ; pulmonary irritation	Chest x ray and pulmonary function testing required

Chemical	NIOSH REL's Recommendations	Health Effect	Remarks
1	2	3	4
4-Dimethylamino-asobenzene	Ca Use 29 CFR 1910.1015	Potential for cancer in human ; produced tumors of the liver and bladder in animals	
Dinitro-ortho-cresol	0.2 mg/m^3 TWA	Central nervous system and metabolic effects	Blood and urine monitoring required ; prevent skin and eye contact; possible delayed effects
Dinitro-toluenes	Ca Reduce exposure to lowest feasible level	Potential for cancer in humans ; produced tumors of the liver,skin,and kidneys in animals ; reproductive system effects	Prevent skin contact
Dioxane	Ca 1 ppm (3.6 mg/m^3) ceiling (30 min)	Potential for cancer in humans; produced tumors of liver, lungs, and nasal cavity in animals; effects on liver and kidney	Blood and urine testing required ; prevent skin contact
Epichlorohydrin	Ca Occupational exposure to epichlorohydrin to be minimized	Respiratory cancer; mutagenesis; reproductive effects; skin, kidney, liver, and respiratory effects	Prevent skin contact
2-Ethoxy-ethanol (See Glycol ethers)			
Ethyl chloride (see Chloroethane)			
Ethylene dibromide	Ca 0.045 ppm (0.38 mg/m^3), 8-hr TWA; 0.13 ppm (1 mg/m^3) ceiling (15 min)	Potential for cancer in humans; mutagenesis; damage to skin, eyes, cardiovascular, liver, spleen, reproductive, respiratory, and central nervous systems	Workers to be warned of potential for reproductive abnormalities and cancer;; hazardous liquid; prevent skin contact.

Chemical	NIOSH REL's Recommendations	Health Effect	Remarks
1	2	3	4
Ethylens dichloride	Ca 1 ppm (4 mg/m^3) TWA; 2ppm (8 mg/m^3) ceiling (15 min)	Potential for cancer in humans; nervous system respiratory, cardiovascular, and liver effects	Nursing infants of exposed mothers at risk.
Ethyleneimine	Ca Use 29 CFR 1910.1012	Potential for cancer in human; produced tumors of the liver and lung in animals	Stringent workplace controls and medical surveillance required
Ethylende oxide	Ca 5 ppm (9 mg/m^3) ceiling (10 min/day); <0.1 ppm (0.18 mg m^3,,8-hr TWA	Peritoneal cancer; leukemia; mutagenesis ; reproductive effects	Blood monitoring and medical counseling recommended
Ethylene thiourea	Ca Should be used in encapsulated form in industry; worker exposure to be minimized	Potential for cancer and teratogenicity in humans; produced tumors of the liver,thyroid,and lymphatic systems in animals	Workers to be informed of carcinogenic and teratogenic hazards ; special attention to be given to thyroid function tests
Fluorides inorganic	2.5 mg F/m^3 TWA	Kidney and bone effects	Urine monitoring required
Fluorocarbon polymers,decomposition products	Various recommendations emphasizing good work practices, engineering controls, and medical surveillance	Lung effects ; polymer fume fever	Workroom air to be monitored for inorganic fluorides and hydrogen fluoride
Formaldehyde	Ca 0.1 ppm ceiling (15 min); represents the lowest reliably quantifiable concentration	Potential for cancer in humans; produced tumors of the nasal cavity in animals	Medical surveillance ; skin protection
Foundries	Various recommendations emphasizing good work practices engineering controls, and medical surveillance	Cancer; respiratory diseases; heat-induced illness; noise-induced hearing loss; vibration-induced disorders; eye injuries ; traumatic and ergonomic injuries	Recommundations limited to foundries that pour molten metal into sand molds
Furfuryi	50 ppm (200 mg/m^3) TWA	Respiratory effects	

Chemical	NIOSH		
	REL's Recommendations	Health Effect	Remarks
1	2	3	4
Glycidylethers	All are ceiling values. (15 min) in ppm (mg/m^3); AGE : 9.6(45); BGE : 4.4 (30) ; DGE : 0.2 (1) Ca ; IGE : 50 (240); PGE : 1 (5)	DGE : Potential for cancer in humans ; produced skin tumors in animals; DGE and other glycidyl ethers : skin and mueous membrane effects ; sensitization potential ; possible hematopoietic and reproductive system effects	Possible additive effects with mixture; medical surveillance
Glycol ethers	Reduce exposure to lowest feasible level	Reproductive effects ; teratogenicity	Prevent skin contact
Hexachloroethane	Ca Reduce exposure to lowest feasible level	Potential for cancer in humans ; produced liver tumors in animals	
Hot environments	Sliding scale limits based upon environmental and metabolic heat loads	Heat-induced illnesses	Recommendations include acclimatization, strict work practices, protective equipment
Hydrazines	Ca All are ceiling values (120 min) in ppm (mg/m^3) : hydrazine : 0.03 (0.04) ; 1, 1-dimethylhydrazine 0.06 (0.15); phenylhydrazine : 0.14 (0.6) ; methylhydrazine : 0.04 (0.08)	Potential for cancer in humans; produced tumors of the lung,liver,blood vessels,and intestines in animals; blood,liver,and skin effects	Blood and urine monitoring and chest x ray required; bowel examination for workers oer age 40
Hydrogen cyanide and cyanide salts	4.7 ppm (5 mg CN/m^3) ceiling (10 min)	Thyroid,blood,respiratory system effects	Concurrent measurement required for HCN when measuring for cyanide salt; trained first-aid personnel and first-aid kits to be available during use; prevent skin and eye contact

Chemical	NIOSH		
	REL's Recommendations	Health Effect	Remarks
1	2	3	4
Hydrogen fluoride	3 ppm (2.5 mg F/m^3) TWA ; 6ppm (5.0 mg F/m^3) ceiling (15 min)	Skin,eye,and airway irritation; bone effects	Pelvic x ray (male workers only) and urine testing required
Hydrogen sulfide	10 ppm (15 mg/ m^3) ceiling (10 min)	Irritation ; severe acute effects involving nervous and respiratory systems	Continuous monitoring required if potential exists for exposure to ≥ 70 mg/m^3 (47 ppm); evacuation required at this level
Hydroquinone	0.44 ppm (2 mg/m^3) ceiling (15 min)	Eye and skin effects	Special provisions for darkroom use
Isopropyl alcohol	400 ppm (984 mg/m^3) TWA ; 800 ppm (1,968 mg/m^3) ceiling (15 min)	Mucous membrane irritation ; possible cancer threat in manufacturing process	Stringent work practices and medical surveillance for manufacturing workers required
Kepone	Ca 1 μ/g^3 TWA	Liver cancer ; nervous system effects	Liver function testing required
Ketones	All are TWA values in ppm (mg/m^3): aetone: 250 (590): methyl ethjyl ketone: 200 (590): methyl *n*-Oropyl ketone: 150 (530); methyl *n*-butyl ketone: 1 (4): methyl *n*-amyl ketone: 100 (465); methyl isobutyl ketone 50(200); methyl iosoamyl ketone: 50 (140); cyclohexanone: 25 (100); mesityloxide: 10(40): diacetone alcohol: 50(240); isophorone: 4(34)	Irritation; liver,kidney,and nervous system effects	Urinalysis required; workers exposed to methyl *n*-butyl ketone to be warned of nervous system effects
Lead,inorganic	< 100 μ*g* pb/m^3 TWA; air level to be maintained so that worker blood lead remains ≤ 60μg/100*g* .	Kidney,blood,and nervous system effects	Blood monitoring required
	Extensive work-practice and personal protection recommendations	Primarily trauma and falls	Tetanus toxoid inoculations and first-aid programs to be instituted

Chemical	NIOSH REL's Recommendations	Health Effect	Remarks
1	2	3	4
Malathion	15 mg/m^3 TWA	Nervous system effects	prevent skin contact; blood monitoring required
Mercury,inorganic	0.05 mg Hg/m^3. 8-hr TWA	Central nervous system and mental effects	Work practices, sanitation, monitoring, and medical surveillance emphasized
2-Methoxy-ethanol (see Glycol ethers)			
Methyl alcohol	200 ppm (262 mg/m^3) TWA: 800 ppm (1 048 mg/m^3) ceiling (15 min)	Blindness; metabolic acidosis	
Methyl chloro-methyl ether	Ca Use 29 CFR 1910. 1006	Lung cancer	
4,4'- Methylenebis-(2-chloroniline)	Ca 3 $\mu g/m^3$ TWA (lowest detectable limit)	Potential for cancer in humans; produced liver and lung tumors in animals	Chest *x* ray; blood and urine testing required
Methylene chloride	Ca Reduce exposure to lowest feasible limit	Potential for cancer in humans; produced tumors of the lung,liver,salivary,and mammary glands in animals	
4,4'-Methylenedia niline	Ca Reduce exposure to lowest feasible limit	Bladder cancer; skin and liver effects	Prevent skin contact
Methyl parathion	0.2 mg/m^3 TWA	Central nervous system effects	Prevent skin contact; blood monitoring required
Monohalomethanes	Ca Exposure to methyl chloride methyl bromide,and methyl iodide should be reduced to lowest feasible level	Potential for cancer in humans; produced tumors of the kidney,forestomach,and lung in animals; methyl chloride should also be considered a potential teratogen	

Chemical	REL's Recommendations	NIOSH Health Effect	Remarks
1	2	3	4
alpha-Naphthyl-amine	Ca Use 29 CFR 1910. 1004	Bladder cancer	
Niax catalyst ESN		Urological disordders;neverous system effects	Work-practice and engineering controls to reduce exposure
Nickel carbonyl	Ca 1 ppd (7 $\mu g/m^3$) TWA (lowest detectable level)	Lung and nassal cancer	Chest x -ray
Nickel, inorganic compounds	Ca 15 µg Ni m^3 TWA	Lung and nasal cancer: skin effects	Chest x ray and pulmonary function testing equired
Nitric acid	2 ppm (5 mg/m^3) TWA	Dental erosion; nasal/lung irritation	prevent skin and eye contact: chest x ray required
Nitriles	All are TWA values in ppm (mg/m^3). acetonitrile: 20 (34); n-butyronitrile; 8 (22); isobutyronitrile; 8 (22); propionitrile; 6 (14); malnonitrile; 3(8); adiponitrile; 4 (18); succinonitrile: 6 (20). All ceiling values (15 min) in ppm (mg/m^3); aetone cyanohydrin: 1 (4); glycolonitrile: 2 (5); tetramethyl succinonitrile: 1 (6). When present as mixtures or with other sources of cyanide, exposure to be considered additive and environmental limit to be calculated.	Hepatic,renal,respiratory,cardiovascular,gastrointestinal,and nervous system effects	Chest x ray and pulmonary function testing required; trained personnel and first-aid kits to be available during use; prevent skin and eye contact.
4-Nitrobiphenyl	Ca Use 29 CFR 1910 . 1003	Potential for cancer in humans; produced bladder tumors in animals	

Chemical	NIOSH		
	REL's Recommendations	Health Effect	Remarks
1	2	3	4
Nitrogen, oxides	NO_2: 1 ppm (1.8 mg/m^3) ceiling (15 min); No; 25 ppm (30 mg/m^3) TWA	Respiratory and blood effects	Pulmonary function testing required
Nitroglycerin and ethyleneglycol dinitrate	0.1 mg/m^3 ceiling (20 min) recommended limit for either substance above or mixtures	Circulatory system effects	Prevent skin contact
2-Nitro-naphthalene	Ca Reduce exposure to lowest feast feasible level	Bladder cancer	Compound matabolizes to beta-naphthylamine, a known carcinogen
2-Nitropropane	Ca reduce exposure to lowest feasible level	Potential for cancer in humans; produced liver tumours in rats	Medical monitoringwith specific emphasisoin liver function tests
N-Nitrosodimethyl-amine	Ca Use 29 CFR 1910,1016	Potential for cancer in humans; produced tumours of the liver, kidney lung and nasal cavity in animals	
Organotin compounds	0.1 mg sn/m^3 TWA	Eye,skin,liver,nervous system,and cardiovascular effects	Chest x ray,blood and urine monitoring,eye tests,heart examination,and nervous system testing required; prevent skin and eye contact
Paint and allied coating products	Various recommendations for the handling of raw materials and finished products; dispersion of pigment or resin particles; thinning,tinting,and shading; filling; and laboratory functions	Injury and a wide range of toxicities considered	Paint and allied coating products include paints vamishes,lacquers, stains,putties, and paint and vamish removers
Parathion	0.05 mg/m^3 TWA	Nervous system effects	Prevent skin contact: blood monitoring required

Chemical	NIOSH REL's Recommendations	Health Effect	Remarks
1	2	3	4
Pesticides		Wide range of toxicities considered; cancer; nervous and reproductive system effects	Blood monitoring required for some grops; workers to be warned of reproductive effects for some compounds; prevent skin contact
Phenol	5.2 ppm (20 mg/m^3) TWA; 15.6 ppm (60 mg/m^3) ceiling (15 min)	Skin,eye,central nervous system,liver,and kidney effects	Prevent skin and eye contact
Phenyl-beta-naphthylamine	Ca Reduce exposures to lowest feasible level	Bladder cancer	Compound metabolizes to beta-naphthylamine,a known carcinogen
Phosgene	0.1 ppm (0.4 mg/m^3) TWA; 0.2 ppm (0.8 mg/m^3) ceiling (15 min)	Respiratory effects	Pulmonary function testing and chest x ray required
Polychlorinated biphenyls	Ca 1 μ g/m^3 TWA (the minimum reliable detectable concentration using the recommended sampling and analytical methods)	Potential for cancer in humans; produced tumors of the liver pituitary gland an leukemias in animals; skin,liver,and reproductive system effects	Blood testing required female workers of child-bearing age and nursing mothers to be warned of potential adverse effects
Polychlorinated biphenyls (PCB's), potential health hazards from electrical equipment fires or failures	Ca Reduce exposure to lowest feasible limit	Potential for cancer in humans; produced tumors of the liver pituitary glad and leukemaias; skin,liver,and reproductive system effects	Fire-related incidents involving PCB's have resulted in widespread contamination of buildings with PCB's and, in some cases, with PCDF's and PCDD's including TCDD. Emergency response personnel, maintenance or cleanup workers, or building occupants may be exposed to these compounds.

Chemical	NIOSH REL's Recommendations	Health Effect	Remarks
1	2	3	4
Precast concrete products industry, comprehensive safety recommendations	Various recommendations for safe work practices and worker training	Injury and death	Equipment,conditions,and many of the tasks specific to the industry are not covered under the existing regulations
beta-Propiolactone	Ca Use 29 CFR 1910,1013	Potential for cancer in humans; produced tumors of the liver, skin, and stomach in animals	none
Refined petroleum solvents	Kerosene: 100 mg/m^3 TWA; all other solvents: 350 mg/m^3 TWA; 1,800 mg/m^3 ceiling (15 min)	Eye,nose,and throat irriation; dermatitis; nervous system effects	Blood and urine monitoring required; action level for petroleum ether rubber solvent naphtha. 200 mg m^3 TWA : action level for mineral spirits and Stoddard solvent: 350 mg m^3 TWA: action level for kerosene : 100 mg m^3 TWA: prevent skin contact
Silica,crystalline	50 μ/m^3 TWA,respirable free silica	Chronic lung diseas (silicosis)	Chest x-ray,pulmonary function testing required
Sodium hydroxide	2 mg/m^3 ceiling (15 min)	Respiratory irritation	Prevent skin and eye contact
Styrene	50 ppm (213 mg/m^3) TWA. 100 ppm (426 mg/m^3) ceiling	Nervous system effects; eye and respiratory system irritation: reproductive system effects	Prevent skin contact,workers to be warned of possible adverse reproductive effects
Sulfur dioxide	0.50 ppm (1.3 ~ mg/m^3) TWA	Respiratory effects	Pulmonary function testing required
Sulfuric acid	1 mg/m^3 TWA	Pulmonary irritation	Prevent skin and eye contact

Chemical	NIOSH REL's Recommendations	Health Effect	Remarks
1	2	3	4
2, 3, 7, 8-Tetra-chlorodibenzo -p-dioxin (TCDD)	Ca Reduce exposure to lowest fessible level	Potential for cancer in humans; produced tumors at many sites in animals; chlorane	None
1,1,1,2-Tetra-chlor oethane	To be handled in the workplace with caution	Central nervous system effects; possible liver and kidney effects	Exposures should be minimized due to the structural similiarity to the carcinogenic chloroethanes
1,1,2,2-Tetra chloroethane	Ca Reduce exposure to lowest feasible level	Potential for cancer in humans; produced tumors of the liver in animals; liver, gastrointestinal,and nervous system effects	Prevent skin contact: blood monitoring required
Tetrachloroethylene	Ca Minimize workplace exposure levels: limit number of workers exposed	Potential for cancer humans; produced tumors of the liver in animals	None
Thiols: *n*-alkane mono thiols, cyclothexanethiol and benzenethiol	All values in ppm (mg m^3) ceilings (15 min): 1-methanethiol: 0.5 (1.0); 1-ethanethiol: 5.0 (1.3); 1-propanethiol: 0.5 (1.6); 1-butanethiol: 0.5 (1.8); 1-pentanethiol: 0.5 (2.1); 1-hexanethiol: 0.5 (2.4); 1-heptanethiol:0.5 (2.7); 1-octanethiol: 0.5 (3.0); 1-nonanetiol: 0.5 (3.3); 1-decanethiol: 0.5 (3.6);	Irritation : eye, skin, blood, and nervous system effects	Blood and urine monitoring required ; prevent skin contact

Chemical	NIOSH REL's Recommendations	Health Effect	Remarks
1	2	3	4
	1-undecanethiol; 0.5 (3.9); 1-dodecanethiol: 0.5 (4.1) 1-hexadecanethil: 0.5 (5.3); 1-octadecanethiol: 0.5 (5.9); cyclohexanethiol: 0.5 (2.4); benzenethiol: 0.1 (0.5); Mixtures of thiols to be controlled by calculation of equivalent concentrations		
o-Tolidine	Ca 20 µg/m^3 ceiling (60 min)	Bladder cancer; nassal irritation	Urine testing required; quarterly urine monitoring recommended; prevent skin contact
0-Tolidine-based	Ca Should be handled in the workplace with caution; minimize exposures	bladder cancer	Substitute less toxic dyes wherever possible
Toluene	100 ppm (375 mg/m^3) 8-hr TWA; 200 ppm (750 mg/m^3 ceiling (10 min)	Central nervous system depressant	
Toluene diisocyanate	0.005 ppm (-.036 mg/m^3) TWA; 0.02 ppm (0.14 mg/m^3) ceiling (20 min)	Respiratory effects	Chest x-ray, blood tests, pulmonary function testing required
1,1,1-Trichloroeth-ane	350 ppm (1,910 mg/m^3) ceiling (15 min); action level set at 200 ppm (1,091 mg/m^3) TWA	Central nervous system liver, and ardiovascular effects	Medical warning of possible congenital abnormalities required
1,1,2-Tri-chloroeth-ane	Ca Reduce exposure to lowest fessible level	Potential for cancer in humans; produced lived tumors in animals; central nervous system effects	

Chemical	NIOSH REL's Recommendations	Health Effect	Remarks
1	2	3	4
Trichloroethylene	Ca 25 ppm TWA	Potential for cancer in humans; produced liver tumors in animals; central nervous system effects	Workers to be warned of hazards; 25 ppm level can be achieved by use of existing engineering control technology
Trimellitic anhydride	Should be handled in workplace as an extremely toxic substance	Pulmonary edema; immunologic sensitization; irritation of pulmonary tract,eyes,nose,and skin	Limit exposure to as few workers as possible while minimizing workplace levels
Tungsten and cemented tungsten carbide	Insolubte tungsten: 5 mg W/m^3 TWA; soluble tungsten: 1 mg W/m^3 TWA; dust of cemented tungsten carbide containing > 2% cobalt): 0.1 mg Co/m^3 TWA; dust of cemented tungsten carbide (containing > 0.3% nickel). 15μ g Ni/m^2 TWA	Lung and skin effects	Pulmonary function testing and chest x ray required
Ultraviolet radiation	For spectral region of 315-400 nm: 1.0 mW/cm^2 for periods > 1.000 sec; for periods ≤ 1,000 sec, 1,000 mW. Sec/cm^3 (1.0 J/cm^3): for spectral region 200-315 nm consult criteria documen	tSkin and eye effects	
Vanadium	Vanadium compounds 0.05 mg V/m^3 ceiling (15 min); metallic vanadium and vanadium carbide 1 mg V/m^3 TWA	Eye, skin and lung effects	Pulmonary function testing and chest x ray required
Vibration syndrome	Jobs should be redesigned to minimize the use of vibrating handtools; powered handtools should be redesigned to minimize vibration	Vibration syndrome; adverse circulatory and neural effects in the fingers	
Vinyl acetate	4 ppm 915 mg/m^3) ceiling (15 min)	Irritation	

Chemical	NIOSH REL's Recommendations	Health Effect	Remarks
1	2	3	4
Vinyl chloride	Ca Lowest reliably detectable level	Liver cancer	Liver function required
Vinyl halides	Ca Vinyl halides to be controlled as specified for vinyl chloride in 29 CFT 1910. 1017 with eventual goal of zero exposure	Liver cancer for vinyl chloride; potential for cancer from the other vinyl halides that have produced liver and kidney tumors in animals	Vinyl halides include viny chloride, vinylidene chloride, vinyl bromide, vinyl fluoride, and vinylidene fluoride monomers
Waste anesthetic gases and vapours	Halogenated anesthetic agents: 2 ppm ceiling (1 hr); nitrous oxide: 25 ppm TWA during periods of use	Reproductive system effects and audiovisual performance decrements	Workers to be advised of potential effects; abnormal outcome of pregnancies of workers and spouses to be documented
Xylene	100 ppm (434 mg/m^3) TWA; 200 ppm (868 mg/m^3) ceiling (10 min)	Central nervous system depressant; respiratory irritation	
Zinc oxide	5 mg ZnO/m^3 TWA: 15 mg ZnO/m^3 ceiling (15 min)	Metal fume fever	

Appendix –B

Clinical Effects of Exposure to Chemicals

Chemical exposures may result from an accident an explosion, a fire, or a spill. A leaking valve, vessel, or pipe, may also contaminate the atmosphere. An exhaust fan may have failed, or the cover of a tank may be damaged or carelessly replaced. Clothing may be contaminated from a dripping overhead line, and walking through a puddle of a spilled chemical may contaminate shoes. Exposures often occur during maintenance work on equipment or when an unfamiliar chemical is introduced into the laboratory or workshop.

The first warning sign of toxic effects from a hazardous exposure is aberrant behavior if the chemical has narcotic properties. Pallor or cyanosis may appear before the victim is aware of overexposure.

The effects of various chemicals are listed in Tables (1-11) that follows.

Table 1. Chemicals that Cause Mild Irritation of Eye and Mucous Membronos

Acetadehyde	*sec*-Amyl acetate
Accetic acid	Arsenic and compound
Acetic anhydride	Benzoyl peroxide
Acetone	Benzyl chloride
Acetylene dichloride	Bromoform
Allyl alcohol	1,3-Butadine
Allyl chloride	*n*-Butyl acetate
Allyl propyl disulfide	*sec*-Butyl acetate
n-Amyl acetate	*tert*-Butyl acetate

sec-Butyl alcohol
n-Butylamine
Butyl cellosolve
n-Butyl glycidyl ether
p-tert-Butyltoluene
Calcium oxide
Chloroacetaldehyde
α-Chloroacetophenone
Chlorobenzene
o-Chlorobenzylidene malononitrile
Chlorobromomethane
Chloroprene
Copper dusts and mists
Copper fume
Crotonaldehyde
Cumene
Cyclohexane
Cyclohexanol
Cyclohexanone
Cyclohexene
Cyclopentadiene
Diacetone alcohol
Dibutyl phosphate
Dibutyl phthalate
o-DichlorobenzeneAccetic
p-Dichlorobanzene
1,3,-Dichloro-5,5-dimethylhydantion
Diethylamine
Diethylaminoethanol
Difluorodibromomethane
Diglycidyl ether
Diisobutyl ketone
Dimethyl phthalate
Dioctyl phthalate
Dioxane
Diphenyl
Dipropylene glycol methyl ether
Epichlorphyddrin
Epoxy resins
2-Ethoxyethanol
2-Ethoxyethyl acetate
Ethyl acetate
Ethyl acrylate
Ethyl alcohol
Ethylamine
Ethyl amyl ketone
Ethylbenzene
Ethyl bromide
Ethyl butyl ketone
Ethyl chloride
Ethyleneddiamine
Ethylene dibromide
Ethyl ether
Ethyl formate
n-Ethylmorpholine
Ethyl silicate
Ferbam
Ferrovanadium dust
Fluoride
Formaldehyde
Formic acid
Furfural
Furfural alcohol
Glycidol
n-Hexane
Hexachloroethane
sec-Hexyl acetate
Hydrazine
Hydrogen bromide
Hydrogen peroxide, 90%
Hydrogen selenide
Isoamyl acetate
Isoamyl alcohol
Isobutyl acetate

Isophorone
Isopropyl acetate
Isopropyl alcohol
Isopropylamine
Isopropyl ether
Isopropyl glycidyl ether
Lithium hyddride
Magnesium oxide fume
Mesityl oxide
Methyl acetate
Methyl acrylate
Methylal
Methylamine
Methyl amy ketone
Methyl butyl ketone
Methyl cellosolve
Methyl cellosolve acetate
Methylcyclohexanol
o-Methylcyclohexanone
Methyl ethyl ketone
Methyl formate
Mmethyl isobutyl carbinol
Methyl isobutyl ketone
Methyl methacrylate
Methyl propyl ketone
α-Methyl styrene
Morpholine
Naphtha, coal tar
Naphtha, petroleum distillates
Nitroethane
1-Nitropropane
Octane
Osmium tetroxide
Oxalic acid
Pentachlorophenol
Pentane
Phenol
Phenyl ether-biphenyl mixture
Phenyl glycidyl ether
Phosphoric acid
Phosphorus (yellow)
Phosphorus Pentachloride
Phosphorus pentasulfide
Phosphorus trichloride
Propyl acetate
n-Propyl alcohol
Propylene dichloride
Propylene imine
propylene oxide
Pyridine
Selenium compounds
Silver, metal and soluble compounds
Sodium hydroxide
Stoddard solvent
Styrene
Sulfur monochloride
Sulfuryl fluoride
Terphenyls
Tetrachloroethylene
Tetrahydrofuran
Tetranitromethane
Thiram
Tin, inorganic compounds
1,1,2-Trichloroethane
Trichloroethylene
1,2,3-Trichloropropane
1,1,2-Trichloro-1,2,2,-trifluoroethane
Triethylamine
Turpentine
Vinyltoluene
Xylene

Table 2. Chemicals that cause Severe Pulmonary irritations.

Acrolein
Ammonia
Antimony
ANTU
Beryllium and beryllium compounds
Boron trifluoride
Bromine
Butyl mercaptan
Cadmium dust/fume
Chlorine
Chlorine dioxide
Chlorine trifluoride
1-Chloro-1-nitropropane
Chloropicrin
Chromic acid and chromates
Chromium, metal and insoluble salts
Cotton dust, raw
Diazomethane
Diborane
1,1-Dichloro-1-nitroethane
Dichloroethyl ether
Diisopropylamine
Dimethylamine
Dimethyl surfate
Ethamnolamine
Ethylene chlorohydrin
Ethyleneimine
Ethylene oxide
Ethyl mercaptan
Fluorine
Hydrogen chloride
Hydrogen fluoride
Hydrogen sulfide
Iodine
Ketone
Maleioc anhydride
Methyl bromide
Methylene bisphenyl isocyanate
Methyl iodide
Methyl mercaptan
Nickelcarbonyl
Nitric acid
Nitroethane
Nitrogen dioxide
2-Nitropropane
Oxygen difluoride
Ozone
Paraquat
Perchloromethyl mercaptan
Perchloryl fluoride
Phosgene
Phosphine
Phosphorus trichloride
Phthalic anhydride
Selenium hexafluoride
Silicone tetrafluoride
Sulfur dioxide
Sulfuric acid
Sulfur pentafluoride
Tellurium hexafluoride
Toluene-2, 4-diisocyanate
Tributyl phosphate
Uranium (natural), soluble and insoluble compounds
Vanadium pentoxide
Zinc chloride fume

TABLE 3. Chemicals that are Pulmonary Sensitizers

Castor bean pomace
Cobalt, metal fume and dust
Enzymatic detergents
Grain dusts
Maleic anhydride
Methylene bisphenyl isocyanate
Methyl isocyanate
Nickel, metal
1-Phenylenediamine
Phthalic anhydride
Platinum salts
Polyvinyl chloride (fume from heated film: meat wrapper's asthma)
Toluene 2,4-diisocyanate
Tungsten carbide
Western red cedar dust
Wood pulp dust

TABLE 4. Agents Causing Benign Pneumoconiosis

Barium and compounds
Graphite, natural
Hematite
Iron oxide fune
Kaolin
Mica
Silica, amorphous
Soapstone
Stannic oxide
Talc, nonasbestos form

TABLE 5. Chemicals that are Pulmonary Carcinogens

Arsenic and compounds
Asbestos
Bis (chloromethyl) ether
Chloromates (see Chromic acid and Chromates)

TABLE 6. Chemicals that Inhibit Cholinesterase

Azionphosmethyl
Carbaryl
Demeton
Dichlorvos
EPN
Malathion
Methyl parathion
Mevinphos
Naled
Parathion
Ronnel
Tetraethyl dithiono-pyrophosphate
Tetraethyl
pyrophosphate

TABLE 7. Chemicals that act as Asphyxiants

CHEMICAL ASPHYXIANTS

Accetonitrile
Acrylonitrile
Cabon monoxide
Cyanide (alkali)
Hydrogen cyanide

SIMPLE ASPHYXIANTS

Acetylene
Argon,neon and helium
carbon dioxide
Dichloromonofluoromethane
Dichlorotetrafluoroethane
Ethane
Ethylene
Hydrogen
Liquid petroleum gas
Methane
Nitrogen
Propane
Propylene

TABLE 8. Chemicals that act as Central Nervous System Depressants

Acetaldehyde
Acetone
Acetylene dichloride
Acrylamide
Allyl glycidyl ether
n-Amyl acetate
sec-Amyl acetate
Benzene
Bromoform
1,3-Butadiene
n-Butyl acetate
sec-Butyl acetate
n-Butyl alcohol
sec-Butyl alcohol
tert-Butyl alcohol
n-Butyl glycidyl ether
Butyl mercaptan
Carbon disulfide
Carbon tetrachloride
Chlorobenzene
Chloromomethane
Chloroform
Cresol, all isomers
Cumene
Cyclohexane
Cyclohexanone
Cyclohexene
Decaborane
Diacetone alcohol
Dichlorodifluoromethane
Dichloroethyl ether
Difluorodibromomethane
Diglyciddyl ether
Diisobutyl ketone
Dipropylene glycol methyl ether
2-Ethoxyethyl acetate
Ethyl acetate
Ethyl alcohol
Ethyl amyl ketone
Ethylbenzene
Ethyl bromide
Ethyl butyl ketone
Ethyl chloride

Ethylene dibromide
Ethylene dichloride
Ethyl oxide
Ethyl ether
Etthyl formate
Ethyl mercaptan
Ethylidene chloride
Furfuryl alcohol
Glycidol
n-Heptane
Hexachloroethane
n-Hexane
see-Hexyl acetate
Isoamyl acetate
Isomyl alcohol
Isobutyl acetate
Isobutyl alcohol
Isopropyl acetate
Isopropyl alcohol
Isopropyl ether
Mesityl oxide
Methyl acetate
Methyl accetylene, propadine mixture
Methylal
Methyl amyl ketone
Methyl butyl ketone
Methyl cellosolve acetate
Methylcyclohexane
Methylcyclohexanol
o-Methylcyclohexanone
Methylene chloride
Methyl ethyl ketone
Methyl formate
Methyl iodide
Methyl isobutyl carbinol
Methyl isobutyl ketone
Methyl mercaptan
Methyl propyl ketone
α-Methyl styrene
naphtha, coal tar
Naphtha, petroleum distillates
Nitroethane
Octane
Pentaborne
Pentane
Phenyl glycidyl ether
Propyl acetate
n-Propyl alcohol
propylene dichloride
Propylene oxide
Pyridine
Stoddard solvent
Styrene
Sulfuryl fluoride
1,1,1,2-Tetrachloro-2,2-difluoroethane
1,1,2,2-Tetrachloro-1,2-difluoroethane
Tetrachloroethane
Tetrachloroethylene
Tetrahydrofuran
Toluene
1,1,1-Trichloroethane
1,1,2-Trichloroethane
Trichloroethylene
Trichlorofluoromethane
1,2,3-Trichloropropane
1,1,2-Trichloro-1,2,2-trifluoroethane
Turpentine
Vinyltoluene
Xylene

TABLE 9. Chemicals that Cause Convulsions

Aldrin
2-Aminopyridine
Camphor
Chlorane
Crag herbicide
DDT
Decaborne
2,4-Dichlorophenoxyacetic acid
Dieldrin
1,1-Dimethylhydrazine
Endrin
Heptachlor
Hydrazine
Lindane
Methoxychlor
Methyl bromide
Methyl chloride
Methyl iodide
Methyl mercaptan
Monomethylhydrazine
Nicotine
Nitromethane
Oxalic acid
Pentaborane
Phenol
Rotenone
Sodium fluoroacetate
Strychnine
Tetraethyl lead
Tetramethyl lead
Tetramethylsuccinonitrile
Thallium, soluble components
Toxaphene

TABLE 10. Chemicals which Cause Skin Irritation

Acetaldehyde
Acetic acid
Acetic anhydride
Acetone
Acrolein
Acrylamide
Acrylonitrile
Alloy alcohol
Alloy chloride
Allyl glycidyl ether
Ammonia
n-Amyl acetate
sec-Amyl acetate
Antimony
Arsenic and compounds
Barium and compounds
Benzene
Benzoyl peroxide
Benzyl chloride
Beryllium and beryllium compounds
Boron oxide
Boreon trifluoride
Bromine
n-Butyl acetate
sec-Butyl acetate
n-Butyl alcohol
sec-Butyl alcohol
tert-Butyl alcohol
n-Butylamine
n-Butyl glycidyl ether
Calcium arsenate

Calcium oxide
Carbaryl
Carbon disulfide
Carbon tetrachloride
Chlorinated diphenyl oxide
Chlorine
Chlorine trifluoride
Chloroacetaldehyde
α-Chloroacetophenone
Chlorobenzene
o-Chlorobenzylidene malononitrile
Chlorobromomethane
Chlorodiphenyl, 42% chlorine
Chlorodiphenyl, 54% chlorine
Chlroform
Chloropicrin
Chloroprene
Chromic acid and chromates
Chromium, soluble chromic and chromous salts
Coal tar pitch volatiles
Copper dusts and mists
Crag herbicide
Cresol, all isomers
Crotonalldehyde
Cumene
Cyanides (alkali)
Cyclohexane
Cyclohexanol
Cyclohexanone
Cyclohexene
DDT
Diacetone alcohol
Dibutly phosphate
o-Dichlorobenzene
p-Dichlorobenzene
1,1-Dichloro-1-nitroethane
Diethylamine
Diethylaminoethanol
Diglycidyl ether
Diisoburyl ketone
Dimethylamine
Dimethylformamide
1,1-Dimethylhydrazine
Dimethyl sulfate
Dioxane
Epichlorhydrin
Epichlorhydrin
Epoxy resins
Ethanolamine
2-Ethoxyethanol
2-Ethoxyethyl acetate
Ethyl acetate
Ethyl acrylate
Ethylamine
Ethyl amyl ketone
Ethylbenzene
Ethyl bromide
Ethyl butyl ketone
Ethylenediaminc
Ethylene dibromide
Ethylene dichloride
Ethyleneimine
Ethylene oxide
Ethyl ether
Ethyl formate
Ethylidene chloride
Ethyl silicate
Fluoride
Fluorine
Formaldehyde
Formic acid
Furfural
Glycidol
n-Heptane

Hexachloronaphthalene
n-Hexane
Hydrazine
Hydrogen bromide
Hydrogen fluoride
Hydrogen peroxide, 90%
Iodine
Isoamyl acetate
Isobutyl alcohol
Isophorone
Isopropylamine
Isopropyl ether
Isopropyl glycidyl ether
Ketene
Lead arsenate
Lithium hydride
Maleic anhydride
Mercury
Mercury, alkyl compounds
Mesityl oxide
Methyl acetate
Methyl acrylate
Methyl acrylate
Methylal
Methyl alcohol
Methylamine
Methyl bromide
Methyl butyl ketone
Methyl chloride
Methycyclohexane
Methylcyclohexanol
o-Methylcyclohexanone
Methylenenbisphenyl-isocyanate
Methylene chloride
Methyl ethyl ketone
Methyl iodide
Methyl isobutyl ketone
Methyl isocyanate
Methyl methacrylate
Methyl propyl ketone
α-Methyl styrene
Monomethylhydrazine
Morpholine
Naled
Naphta, coal tatnaphtha, petroleum distillates
Nickel, metal
Nitric acid
Nitrobenzene
Nitroethane
Nitromethane
Octachloronaphthalene
Octane
Osmium tetroxide
Oxalic acid
Pentaborane
Pentachloronaphthalene
Pentachlorophenol
Pentane
Perchloromethyl mercaptan
Phenol
p-Phenylenediamine
Phenyl ether
Phenyl ether-biphenyl mixture
Phenyl glycidyl ether
Phenyl glycidyl ether
Phenylhydrazine
Phosgene
Phosphoric acid
Phosphorus (yellow)
Phosphorious pentachloride
Phosphorus pentasulfide
Phosphorus trichloride
Phthalic anhydride
Phthalic anhydride
Picric acid

Platinum, soluble salts
Portland cement
Propyl acetate
n-Propyl alcohol
Propylene dichloride
Propylene imine
Propylene oxide
Pyrethrum
Pyridine
Quinone
Rotenone
Selenium compounds
Silver, metal and soluble compounds
Sodium hydroxide
Stoddard solvent
Styrene
Sulfur dioxide
Sulfur dioxide
Sulfur monochloride
Sulfuric acid
Tellurium
Terphenyls
Tetrachloroethane
Tetrachloroethylene
Tetrachloronaphthalene
Tetranitromethane
Tetryl
Thiram
Tin, organic and inorganic compounds
Toluene
Toluene 2,4-diisocyanate
o-Toluidine
Toxaphene
Tributyl phosphate
1,1,1-Trichloroethane
Trichloroethylene
Trichloronaphtalene
2,4,5-Trichlorophenoxyacetic acid
1,2,3-Trichloropropane
1,1,2-Trichloro-1,2,2
trifluoroethane Triethylamine
2.4.6-Trinitrotoluene
Turpentine
Uranium (natural), soluble an insoluble compounds
Vanadium pentoxide
Vinyltoluene
Xylene
Zinc chloride fume

TABLE 11. Chemicals that act as Skin Sensitizers

Acetic anhydride
Allyl glycidyl ether
Benzoyl peroxide
Beryllium and beryllium compounds (granuloma)
n-Butyl glycidyl ether
Chromic acid and chromates
Cobalt, metal fume and dust
Cresol, all isomers
o-Dichlorobenzene
Dintrochlorobenzene
Epoxy resins
Ethyl acrylate
Ethylenediamine
Formaldehyde
Iodine
Isopropyl glycidyl ether
Maleic anhydride
Mercury
Naphthalene

Nickel, metal
p-Phenylenediamine
Phenyl glycidyl ether
Phenylhydrazine
Phthalic anhydride
Picric acid
Platinum, soluble salts
Tetryl
Thiram
Toluene 2,4-disocyanate
2,4,6-Trinitrotoluene
Vanadium pentoxide